The Microbiome, Gut Health and Oriental Medicine

of related interest

Psycho-Emotional Pain and the Eight Extraordinary Vessels
Yvonne R. Farrell
Foreword by David Chan
ISBN 978 1 84819 292 8
eISBN 978 0 85701 239 5

Acupuncture for Surviving Adversity
Acts of Self-Preservation
Yvonne R. Farrell
Foreword by Russell Brown
ISBN 978 1 78775 384 6
eISBN 978 1 78775 385 3

Treating Children with Chinese Dietary Therapy
Sandra Robertson
Foreword by Lillian Bridges
ISBN 978 1 78775 318 1
eISBN 978 1 78775 319 8

The Acupuncture Point Functions Charts and Workbook
Erica Joy Siegel
ISBN 978 0 85701 390 3
eISBN 978 1 78775 009 8

The Microbiome, Gut Health and Oriental Medicine

An Integrated Approach

Lisa Lee

SINGING DRAGON
LONDON AND PHILADELPHIA

First published in Great Britain in 2022 by Singing Dragon,
an imprint of Jessica Kingsley Publishers
An imprint of Hodder & Stoughton Ltd
An Hachette UK Company

1

A CIP catalogue record for this title is available from the British Library and the Library of Congress

ISBN 978 1 78775 985 5
eISBN 978 1 78775 986 2

Printed and bound in Great Britain by CPI Group

Jessica Kingsley Publishers' policy is to use papers that are natural, renewable and recyclable
products and made from wood grown in sustainable forests. The logging and manufacturing
processes are expected to conform to the environmental regulations of the country of origin.

Jessica Kingsley Publishers
Carmelite House
50 Victoria Embankment
London EC4Y 0DZ

www.singingdragon.com

To Louis

Acknowledgments

This book would not have come into existence without the contributions, both direct and indirect, of many people, and special thanks go to them. In particular, I want to mention the many patients who, each in their own way, provided the nuggets of knowledge and experience that gave me the opportunity to learn about, reflect on and explore the ideas presented in the book and whose healing journeys brought to life the teachings of Oriental Medicine. I am immensely grateful to Claire Wilson at Singing Dragon for giving me the chance to take forward this project and for providing the space and structure I needed to string together what were initially just some loose ideas. I want to also acknowledge the remarkable acupuncturists and teachers I have been fortunate to meet and listen to over the years, and my wonderful family, near and far.

Contents

Preface

When we look back at key moments in our lives, we are often surprised to see how both purposeful and chance events have come together at precise points in time to get us to where we are. Never was this truer than for the story behind this book and the subject of the book itself—a story of different pieces of a puzzle coming together over time to create a complex yet intriguing picture of the potential that exists within every one of us to create, achieve and maintain good physical and mental health. The potential I am talking about is the microbiome.

The microbiome is a world in its own right, populated by trillions of different bacteria, fungi, viruses and other micro-organisms, yet found across and within every one of us. Microbiome research is at the very forefront of some of the latest health developments and scientific research across wide-ranging disciplines, including nutrition, psychiatry, medicine, environmental sciences and anthropology, to name a few, precisely because of the microbiome's significance for the living world. The microbiome is essentially about life and the infinite possibilities this affords, making it a concept that to all intents and purposes is a mirror image of the workings of Oriental Medicine. It is because of this that the theory and practice pertaining to the microbiome brings with it also the opportunity to delineate a common ground between Eastern and Western medicine.

Like many people probably, I had been for most of my life only vaguely aware of the presence and role of microbes in our lives. For many years, I had in fact pictured the microbial world as a simple duality, with two distinct camps of microbes—the good and bad bugs—one of which was found in my yoghurt and the other on unwashed hands. This simplistic understanding first started shifting about 16 years ago when, through a series of encounters with different healthcare professionals following a fatal case of sepsis in my own family, I realized that the microbial world we are part of, and which is part of us, often hangs precariously, relying on a delicate and still not fully understood balance. This shift in my thinking also led to the realization that the power of microbes is immense, even though by and large we are completely unaware of their presence as they quietly co-exist with us. I did wonder, given the apparent fragility of this partnership between host and microbe, how any kind of balance is so effectively maintained most of the time and, as a result, is able to provide what is for many a taken-for-granted state of health. And under what conditions does this alliance and the peaceful co-habitation of micro-organisms break down?

I put these questions to one side and was suddenly brought back to thinking about them some years later, when working as an acupuncturist with patients who presented with a range of different health conditions, both physical and mental, that somehow all seemed to have a particular pattern to them. Practitioners of Oriental Medicine are accustomed to hearing patients describe what may seem like an "odd" collection of symptoms, yet which, in the context of a Traditional Diagnosis, such as the one undertaken in an acupuncture treatment, makes perfect sense. What stood out in these patients, however, was the consistency with which chronic gastro-intestinal health issues, skin problems, anxiety, poor sleep and dietary difficulties repeatedly intersected in the patients' medical history. These patients had suffered for years with symptoms that were having a significant impact on their everyday life, and yet in almost all cases their doctors had been struggling to treat them or had provided what patients felt were unsatisfactory answers on the causes and treatment options. A diagnosis of irritable bowel syndrome (IBS) was commonly made and left at that, while some patients were simply told that their health problems were due to a past parasitic infection, chronic fatigue or "something else." As the distressing gastro-intestinal symptoms had come hand in hand with various degrees of anxiety, it had become hard for patients to dissociate the two disorders, as they had come to experience gastro-intestinal problems and anxiety both mentally and physically. This twinned manifestation of mental and physical health issues is of little surprise to a practitioner of Oriental Medicine. Matters of the body and mind are in this system of medicine seen as intricately related, and physical, mental and, importantly too, spiritual health are therefore treated concurrently. As I was traveling home from my clinic thinking about these patients, I serendipitously caught a snippet of a piece on the radio about how fermented foods positively affect gut health and as a result overall physical health and mental wellbeing. Not only did this resonate strongly with the core principle in Oriental Medicine that "food is the number one medicine" and the long-held view across many traditions of medicine that many, if not all, diseases begin in the gut, it also provided me with a new therapeutic outlook on how to work with these and other patients.

Up until this point in time, the dietary recommendations I had been giving to my patients had been based on those used in Traditional Chinese Medicine (TCM), and which are based on pattern identification and food energetics (e.g. Leggett 2014a, 2014b; Pitchford 2002). As I brought gut health into the diagnostic and therapeutic equation, I found myself increasingly encouraging my patients to read books about gut health (e.g. Collen 2015; Enders 2015; Mosley 2017) to help them fully grasp the impact of diet and lifestyle on the microbial world within, and its link to both physical and mental health. I explained to them that this approach gave an added dimension to the food recommendations made according to the principles of Oriental Medicine. I also started fine-tuning the dietary recommendation I provided to patients to better align their individual acupuncture treatments with the microbiome literature. It was at this point that I realized that it was time to truly bring together these two bodies of knowledge and fields of practice—and so this book was born.

This book is of course first and foremost a story of the meeting of Eastern and Western medicine, something that is much needed for any student or practitioner of Oriental Medicine working within a world dominated by Western lifestyles and diets. Equally, it is hoped that the book offers a new perspective to healthcare providers working within conventional medicine or in the areas of microbiome and gut health. Significantly, given the growing availability of "microbiome-friendly" foods, supplements and even cosmetics, the book provides a timely guide to navigating the complex field of the microbiome to help patients understand through the lens of Oriental Medicine how food and lifestyle choices may influence their health and wellbeing. Bridging traditions in this way involves some degree of "stepping out of the box," and while I hope the book to be in many ways illuminating to readers, its intention is to open up new possibilities rather than refute the validity of different therapeutic and diagnostic approaches, whether rooted in Oriental Medicine or not.

Introduction

The microbiome is proving to be an exciting, albeit convoluted, area of scientific research and medicine, which, over the last two decades, has sparked an unprecedented level of interest in the field. The attention the microbiome has received is certainly justified, given its role and the largely untapped potential it offers in the prevention and treatment of a wide range of diseases and symptoms, both physical and mental. Microbiome science is frequently described as offering a new frontier in medicine, as the application of advanced sequencing methods to study microbes' genetic and metabolic profiles (Marchesi *et al.* 2016), has unearthed the microbiome's remarkable range of functions and scale of influence. This knowledge has allowed scientific communities around the world to refocus attention on better understanding the micro-organisms living inside and on the body, and importantly too on finding ways to utilize them for the maintenance of good health.

The microbial world is revealing a level of complexity, but also of possibilities, of which many of us are living in blissful ignorance. Even scientists and researchers in the field are still trying to come to grips with the manifold, intricate and elaborate symbiotic relationships that exist between microbes and their hosts, and among microbes themselves. The dearth of knowledge in this area is to some extent surprising when we consider how the existence of microbes goes back billions of years and predates that of humans (Yong 2017), and that Chinese Medicine theory itself has for millennia established principles pertaining to the pathogenic effect of micro-organisms such as parasites. Concretely, however, it was not until observations and depictions of living micro-organisms in and around us started to be made in the latter part of the seventeenth century with the advent of the first microscopes, that the scene was set for key developments in microbiome research, including in the nineteenth century, with the pioneering work of the German pediatrician Escherich who studied babies' intestinal flora, and later the French doctor Tissier who developed probiotics to treat diarrhea in infants, or the Russian immunologist Metchnikov who advocated the health benefits of fermented milk (Farré-Maduell and Casals-Pascual 2019).

On reading this, practitioners of Oriental Medicine may consider the microbiome to be an interesting topic which nevertheless firmly belongs to colleagues working within conventional medicine. Meanwhile, healthcare practitioners trained in Western medicine and others with an interest in the microbiome may be wondering how a practice such as acupuncture could be relevant to them. Let us, however, for a moment consider the microbiome from a different angle and ask ourselves whether the microbiome's origin and transformation, and the micro-organisms within, could be part of the story of acupuncture itself. While this may seem at first to be an odd proposition, looking at the theory and application of acupuncture as a reflection of the microbiome's evolution and inner workings provides a useful approach to understanding how Eastern and Western medical perspectives not only complement one another, but are in fact working on very similar assumptions.

As an acupuncturist, I have been struck by the parallels between microbial therapeutic applications and Oriental Medicine. In particular, the idea that humans have a natural in-built ability to heal, that ill health comes from certain imbalances, or that health is enabled through multiple symbiotic relationships which maintain homeostasis, are all principles that are central to the rationale and aims of both acupuncture and microbiome theory and practice. The more one delves into microbiome research, the stronger the echo. Practitioners of the various styles of acupuncture and modalities rooted in the ancient system of Oriental Medicine, including herbs, food therapy, *Tai Chi* or *Qigong*, will immediately see how the key tasks of the microbes to protect, nourish and transform mirror those functions associated with the energetic potential afforded by the *Vital Substance* of *Qi*. The word *Qi* in Chinese means air or breath, but in acupuncture theory it is loosely translated as "life-force," as it performs a wide range of vital tasks across body systems.

Significantly too, Chinese Medicine, which finds its origin in Taoism, is deeply rooted in the world of microbes. According to Taoist myths, humans' first appearance can be traced back to a giant named *Pangu*, whose birth was triggered by the interaction of the opposing forces of *Yin* and *Yang*, and which as a result released the universe to create Heaven and Earth. When the giant died and his body decayed, its different parts formed the natural world as we know it, with mountains, valleys, forests, marshes, rivers and oceans. The parasites which inhabited this rich and diverse land-scape, according to the story, represent the manner in which the first humans came into existence. From this understanding of the origins of life, traditional acupuncture evolved as a practice which recognized humans as fundamentally and inevitably part of the natural world, and which resonates in all aspects of life's workings (Larre and de la Vallée 1995), something that is used both diagnostically and therapeutically within its theoretical models. As such, harmonious living with the natural flow of life is inherently intuitive and a norm that represents a core existentialist principle of *Taoist* ethics. *Pangu*'s story is remarkable as it provides an allegorical depiction of the microbiome, and of the symbiotic relationship between parasite and host. Its description of humans' emergence from the two opposing and yet complementary forces of *Yin* and *Yang* represents an impossible alliance, much like eukaryotes' union of bacteria and archaea which is believed to have enabled all living things to be (Yong 2017).

These fascinating parallels foreground as yet little explored interpretations of microbiome science and Oriental Medicine, and of the cross-over of these two medical traditions, thereby opening the door to therapeutic synergies that can be used by practitioners and patients.

Ancient belief systems and practices such as *Taoism* and acupuncture, which see human health as deeply embedded in the natural world, may seem anachronistic in today's world. However, more than ever perhaps, humans are starting to reawaken to the realization that we are indeed deeply connected to nature. The relevance of such principles is clear when we consider the influence of humans on climate change and the subsequent effects this is having on human health. Environmental changes, and particularly temperature rises linked to human behavior, and the overindulgences and ill treatment to which our planet has been subjected through pollution, over-exploitation and from a lack of respect for nature, are now seen as key vectors of ill health and premature deaths, including from cardiovascular and respiratory conditions, as well as infectious diseases transmitted through insects and animals (World Health Organization 2018). Such patterns are a stark reminder of the close union that exists between humans and nature and illustrate the ecological model that underlies the maintenance of health. They also validate the rationale for nourishing the balance of the microbiome through appropriate food, lifestyle choices and Oriental Medicine, as this seamlessly enables harmonious living with nature and ourselves.

It is in this vein that the book explores the connections between the natural world we inhabit, and health strategies rooted in Oriental Medicine and the microbiome, with a particular emphasis on nutrition and the physical, mental and spiritual parameters of health as understood in acupuncture. More specifically, it is argued that an integrated microbiome–Oriental Medicine therapeutic health model ensures a beneficial alignment of humans and nature by helping recognize the wide-ranging influences on the microbiome, including energetic shifts linked to seasonal or climatic factors. This provides a very simple and yet effective framework for achieving physical health and mental wellbeing, which mirrors basic principles that can be observed in the living world. For instance, we know that flowers grow better within set conditions, such as suitable levels of moisture and sunlight, and that these levels naturally vary across the day, night, time of year and location. If the plant is deprived of this sustenance, there will be no growth, or conversely if it has too much of this nourishment relative to its needs, there may be deleterious effects. We also know that in nature, the potential for repair exists with the right conditions. As we progress through the chapters, we will see how such optimal conditions can be supported through an approach to microbiome health that is guided by Oriental Medicine theory, allowing mind and body to adapt, respond and heal appropriately to achieve homeostasis and the maintenance of good health.

Importantly, the shared ecological models of Oriental Medicine and the microbiome are used to bridge the divide between Eastern and Western systems of medicine, both in theory and in practice. By so doing, it is hoped to highlight the opportunities that exist for these two systems of medicine to learn from each other to support interdisciplinary knowledge in the field, and for practitioners of Oriental Medicine to integrate the existing and growing knowledge of the microbiome into their everyday work. Throughout the chapters, there is a strong emphasis on recognizing and utilizing the synergy between Eastern and Western medicine, rather than advocating the superiority of one system over another. The book therefore draws on a broad body of knowledge to look at the microbiome through the lens of Oriental Medicine: it harnesses insights derived from the fascinating work of microbiologists and other scientists and researchers working in the field of the microbiome, while contextualizing these within theoretical concepts used in the practice of acupuncture.

We start our journey into the world of the human microbiome and Oriental Medicine by exploring in Chapter 1 some of the key features of the microbiome and its effect on overall health. Particular attention is paid to the gut microbiome due the pivotal role it has been found to play in the health and regulation of other microbiomes. From here, we shift the emphasis to seeing how the interpretations of health and disease found in relation to the microbiome fit within the framework of Oriental Medicine. Little attention has been paid to date to establishing meaningful ways to integrate the valuable knowledge emerging from microbiome research into the everyday practice of Oriental Medicine. Chapter 2 is therefore an attempt at defining the convergence of Western and Eastern medicine in relation to the microbiome, providing practitioners with an understanding of how microbiome knowledge can be conceptually interpreted in Oriental Medicine and integrated into their practice.

This exploration is taken to a deeper level in Chapter 3, where the mighty gut microbiome is focused on. The gut microbiome's influence on the microbiome of other organs, sites and many health conditions underlines several axes of health, notably gut–brain and gut–lung. The chapter looks at these relationships and places them in the context of *Five-Element* acupuncture theory, looking at *Elemental* associations and interrelationships between *Organs*. The *Elements—Metal, Water, Wood, Fire and Earth*—are considered important to understand the congruence between the microbiome and acupuncture as they define energetic patterns, interactions and movements of health and disease.

In Chapter 4, we move on to the practical application of understanding, utilizing and influencing the health of the microbiome in both Occidental and Oriental systems of medicine, by integrating microbiome food science with the principles of food energetics. The concept of food energetics is used by practitioners of Oriental Medicine as a framework to establish what foods should be prioritized and when by patients, as it considers individual presentations, their environment and the qualitative effect of foods, notably heating, cooling, drying or moistening. The chapter considers these Chinese Medicine principles in the context of microbiome-friendly foods, providing a fine-grained dietary approach that integrates the therapeutic understanding of food from each system. Chapter 5 builds on this to provide hands-on ideas for food preparation that boost appropriately our microbiome and energies across the seasons by recognizing the *Elemental* affinities of specific dishes.

It is crucial to recognize how lifestyles play a big part in our health and wellbeing in both Oriental Medicine and microbiome theory. Sleep, exercise, fresh air and living connected with the natural rhythms of life and nature are all aspects that are prioritized and treated in therapies such as acupuncture, and which are also considered important to the health of the microbiome. Chapter 6 therefore explores these aspects, yet seeks to do so in a manner that is consistent with the spirit of this book. More specifically, as the ideas presented throughout the book rest upon the notion central to Oriental Medicine that physical health and mental wellbeing are not only intertwined but also are dependent on connections across all aspects of our lives, it is fitting to consider an approach to nourishing the balance of our microbiome as a *Microbiome Way* that focuses on the needs and preferences of every individual rather than a strict dietary or a one-size-fits-all lifestyle approach. The *Microbiome Way*, it is argued, allows for an outlook on everyday wellness which is both empowering and sustainable.

The book concludes by briefly reflecting on the themes and issues discussed throughout the chapters and highlights the importance of finding ways to integrate in a meaningful manner knowledge from both Eastern and Western medicine. It is worth pointing out in this respect that the book does not claim any kind of fundamental truth on the microbiome in Oriental Medicine. The aim is simply to bring together two fields of health research and practice that may allow patients to sustain a happy and healthy life. While some may look doubtfully at the ideas presented in the book, as health practitioners we owe it to our patients to, at the very least, consider the possibilities for good health this synergy may provide. Moreover, nourishing the balance represents a perspective on health and wellbeing that advocates harmony with ourselves and the world we live in, and from this point of view alone, it is hard to argue against at least giving it a try.

Chapter 1

The Making of the Microbiome

What is the microbiome? This may seem to be a straightforward question yet giving a definite and concise answer is far from easy. There is a lot going on inside humans that has still not been fully mapped out, and the world of microbes within us and on us—our microbiome—is a case in point. As is often the case in scientific and academic circles, this "universe within" is itself a hotly debated subject, not least when it comes to naming and defining it (Riccio and Rossano 2020). The terms "microbiome" and "microbiota" are often used interchangeably to refer to this complex internal world, although there is now scientific consensus that the former refers more precisely to the micro-organisms and their genes, and the latter relates to the collection of microbes in a specified environment (Berg *et al.* 2020). This microbial entity is also popularly shortened to "biome" (Mosley 2017), a term which, however, is more commonly used in ecological and environmental sciences to describe the fauna and flora adapted to certain climates, of which ecosystems are a subset. Given that the main aim of this book is to bring together principles of the natural world which imbue Oriental Medicine with those of the human body which dominate Western medicine, the term "micro*biome*" is seen as well suited and allows key aspects of the microbiome to be considered, notably its basic make-up, factors that influence it and how it is linked to wider health patterns.

There is much more than just semantics that make defining the microbiome a challenging endeavor. Part of the problem is that there is still much that remains unknown about the microbiome, despite the great advances that have been made over the last two decades and the vast amount of ongoing research in this area. Of all the different microbiomes that exist in and around the body, the gut microbiome has been the most closely scrutinized, partly due to its size, the variety of species it contains, and the pivotal role it has been found to play in health and disease. Gut microbiome research has helped us understand the connections and communication pathways that exist within and among communities of microbes and the wide-ranging functions they perform, including immune, digestive and metabolic functions. This has resulted in some radical shifts in thinking across different health disciplines which have brought Western medicine much closer to Oriental Medicine's long-held view of health as an integrated mind–body system. The evolutionary entanglement of humans and microbes has made these micro-organisms key players in human health in both systems of medicine. It is helpful to look more closely therefore at the history of the microbiome and scientific developments in the field to understand what is meant by this and the implications for practitioners and patients.

MICROBIAL OMNIPRESENCE

Microbes existed long before humans and over time provided the evolutionary mechanisms that have made humans, plants and other animals what they are today. As pointed out in the Introduction, it is believed that the eukaryotic cells common to all living things were the result of a "chance" fusion of archaeal cells with bacteria (Yong 2017), as a bacterium became trapped within a cell, forming the nucleus where DNA is stored. Interestingly, the mitochondria—the organelles that reside within the cytoplasm of the eukaryote and whose job is to power the cell—have also been found to have their own DNA (Chial and Craig 2008), thus forming a second genome. While different in form and having a much smaller number of base pairs than the human nuclear genome, the mitochondrial cell shares an important characteristic with the eukaryotic cell, as it too is believed to have a bacterial origin, which has seemingly allowed it to play a critical role in inflammatory responses (Meyer *et al.* 2018). Bacteria it appears therefore are more than simply tolerated: they are evolutionary allies and perform vital regulatory functions too, demonstrating the symbiotic nature of the relationship of humans and bacteria.

The notion of microbial symbiosis came to prominence some time ago, notably with the work of renowned French microbiologist Louis Pasteur, whose work on fermentation and the action of microbes led him to develop a simple procedure which became known as pasteurization, whereby pathogens could be eliminated from foods and drinks when these are heated for around 30 minutes above 60 degrees Celsius. While his work on pasteurization allowed him to demonstrate how pathogens can be eradicated, Pasteur also argued that, although in some cases germs are pathogenic, complex forms of life such as humans and animals are dependent on microbes. This theory was pivotal in setting the foundation for the many studies which have found how specially bred germ-free mice have reduced lifespans and increased susceptibility to disease and chronic health problems (Rackaityte and Lynch 2020).

The ubiquity and interactions of microbes with other living organisms underline the ways in which humans both play host to extensive communities of microbes and are actors in the microbial ecosystem. The sheer number of microbes led some scientists to initially conclude that we are perhaps better seen as more microbe than human, as the ratio of microbial cells, and therefore gene count, was believed to outnumber that of human cells (Collen 2015).

This estimate has more recently been revised (Abbott 2016), and it is suggested instead that there is quasi-parity in the average microbial and human cells ratio. It is nevertheless estimated that there are trillions of micro-organisms in each person, and more incredibly still, that the exact profile of each person's microbiome is unique. Each microbiome has its own delicate balance of different types of bacteria, fungi, viruses and other cells, with the bulk being bacteria. Together these micro-organisms maintain good physical and mental health, for instance by helping digest foods and extracting nutrients from food, by metabolizing medication, or by teaching specific response and adaptive measures to the immune systems in order to ward off pathogens and fight illnesses (Harvard T.H. Chan School of Public Health 2021a).

As well as being person specific, the microbiome is also highly dynamic and varies considerably from one part of our own body to another. For example, the skin has around one thousand different species of bacteria living on it (Eisenstein 2020a), with a combination of bacteria that will vary from person to person, and from one person's hand to their other hand. Every time a person's hand is touched or is exposed to a non-native microbiome, for example when gardening, cleaning muddy vegetables or stroking a dog (Yong 2017), the microbial profile may change, even if for a short period of time. Just like in nature, this ecosystem is affected by climatic changes too: warm, damp spaces such as armpits harbor different varieties of bacteria to dry, exposed areas such as hands or forearms. The moist environment of the mouth is host to a multitude of species which interact with other microbiomes (Nelson-Dooley 2019). These localized variations provide conditions that are favorable to the dominant groups of bacteria found in each site. The skin, for instance, is dominated by *Staphylococcus* bacteria, the mouth by *Streptococcus*, the vagina by *Lactobacillus*, the gut by *Bacteroides* (Yong 2017), even if significantly in each of these there are countless other micro-organisms playing their part in ensuring homeostasis within and beyond the site-specific microbiome.

"NOT BUILT IN A DAY"

All great things take time to build, and the microbiome is no exception. When exactly the building work of each person's microbiome starts is itself not entirely clear. The principle that we are born with a tabula rasa in the microbiome—a kind of clean slate of microbes—has for a long time been a commonly held view. More recently, however, it has been suggested that a baby's exposure to certain microbes may start within the amniotic sac, which would hence mark the start of the process of colonization in utero. While this view has been challenged, what is clear is that host genetics do shape the microbiome (Hofer 2019), and as such prenatal predispositions are influential.

The immediate postnatal period is widely considered a critical time for the early establishment of the microbiome (Rackaityte and Lynch 2020), as the birthing process is a key point when the baby's exposure to microbes begins. The presence of microbes during and after birth provides newborns with crucial defenses against pathogens, while regulating inflammatory responses, as well as homeostatic and metabolic functions to support their development. A major influence on the initial colonization occurs during the birth itself, as babies born by caesarean section are first exposed to the many commensal *Staphylococcus* microbes living on the skin, whereas those babies born vaginally receive more *Lactobacillus* species (York 2019). Soon after birth, the baby's diet starts shaping the early establishment of particular communities of microbes. Colostrum—the highly nutritious and concentrated form of breastmilk first produced by the mother—contains a rich cocktail of micro-organisms and probiotics that supports the baby's health and development ex utero. Whether or not the baby is breastfed and for how long will carry on influencing his or her microbiome. Gradually the baby will become exposed to more bacteria through interactions with the living world, whether these are from family members, pets or simply from playing in the garden (Rackaityte and

Lynch 2020). By the time the child turns three, a fully developed microbiome will be established (Eisenstein 2020b) which will have been shaped by nutritional, environmental and other factors, such as early use of antibiotics (Underwood *et al.* 2020).

This early period of life is now recognized as a vital time in defining the make-up of the microbiome, which will enable appropriate responses to acute disease as the child grows and will reduce the likelihood of developing various disorders later in life (Eck *et al.* 2020). The early foundations and lasting influence of the microbiome indicate therefore stability over time and resilience—something that may be seen either positively or negatively depending on what start to life a person has had. Some individuals will have had a less than optimal basis for their microbiome, and this has been correlated with the growing rates of allergies, chronic inflammation and auto-immune disorders that are now increasingly common across the developed world.

WIDENING THE NET

The early years are not everything, however, as disruptions to the microbiome—for instance, after a course of antibiotics or from other medication that interferes with its composition—can have a lasting effect and may even lead to some species never returning (Clutter 2019). What is more, there is evidence that modern urban Western lifestyles change in subtle yet complex ways the balance of the microbiome (Eisenstein 2020b), and that this may explain why rates of asthma and allergy have increased in parallel with sanitization and development. Although it was first believed that this increased hygiene had lowered exposure to pathogens, which in turn had been affecting the composition of the microbiome and thereby immune functions, it is now believed that it is the lack of exposure to commensals that disrupts the symbiotic relationships that lower the incidence and progression of chronic conditions (Bondar 2019).

While a precise causal link between microbiome and health is still not fully established, research in this area is finding associations between a microbiome's profile and the occurrence of specific diseases—such as type 2 diabetes and inflammatory bowel disease—and even preterm labor (Ganguly 2019). It is important to understand also that dysbiosis—an imbalance in the microbial communities—appears to have other indirect effects and may act as a catalyst for wider health issues. For instance, inflammation from microbiome imbalances may be associated with hormonal fluctuations, thus indicating the potential role of microbes to have a knock-on effect on various aspects of the reproductive and endocrine system (Qi *et al.* 2021) and therefore on conditions that are becoming more and more common, such as early puberty, infertility and severe menopausal symptoms. It is worth remembering too that the onset of one health condition is frequently associated with a range of other health problems and the development of comorbidities, as body systems are put under additional strain or compensate for the physiological changes triggered by the first disease. This means there may be further problems linked to both the direct and indirect effects of dysbiosis.

Interestingly, microbes are believed to play a critical role not only in preventative and therapeutic medicine, but also in the management of medication, including side effects, and other medical interventions, such as surgery. In particular, it has been found that micro-organisms are capable of aiding with the metabolization of certain medicines (Weersma, Zhernakova and Fu 2020), suggesting that variations in the effect of certain medicines on different individuals, or even susceptibility to side effects, could be related to individual microbiomes. Questions are now also rightly being asked about post-operative complications, such as wound infection, which is a major risk for most surgical interventions, and possible links to the gut flora (Lederer *et al.* 2017). The wide net cast by the microbiome underlines the scale of its network, and the way in which it may prevent the establishment of a primary condition and inhibit cascading ill health. This understanding may help reduce the incidence of complex and hard to treat conditions. It also demonstrates the range of therapeutic goals that may be achieved through microbiome health management.

A MIND–BODY SYSTEM

The microbiome's nexus of functions across body systems provides the basis for a highly integrated interpretation of human health, which is repositioning healthcare theory and practice by moving away from what are often-siloed medical disciplines. It is especially significant to note how microbiome research has demonstrated that patterns of dysbiosis are not limited to the occurrence and progression of physical ill health, as a tangible association has been found between mental health and gut health (Foster, Rinaman and Cryan 2017) via the gut–brain axis (GBA). Such links mark a momentous shift in thinking that unites disciplines and systems of medicine, and also squarely brings the microbiome into the realm of mind–body therapies, such as acupuncture.

It is worthwhile noting that the basic idea of a gut–brain connection is not novel in itself. The occurrence of a range of gastro-intestinal health problems, both chronic and acute, has for some time already been recognized in patients with conditions including depression, anxiety, autism, bipolar disorder, schizophrenia and Parkinson's (Butler *et al.* 2019). It is also the case that the basic bidirectional communication between gut and brain and the feedback loops that exist are well understood. For example, messages of satiety are under normal conditions sent to the brain after eating, and conversely the brain sends instructions pertaining to gut function, for example to put digestive functions on hold when entering "fight or flight" mode. However, what this more recent knowledge of the trillions of gut micro-organisms is doing is first facilitating both a deeper and more detailed understanding of the complex interactions between gut and brain, and secondly opening up new possibilities for the treatment of various disorders, including psychiatric conditions, in a field known as "psychobiotics" (Pennisi 2020). Work in this area has revealed specific microbial patterns among those suffering with mental health disorders (Butler *et al.* 2019; Johnson 2020) and how disruptions to the microbiome, such as through antibiotics, can be a contributory factor of dysbiosis and the prevalence of mental illness (Rogers *et al.* 2016).

This understanding has earned the microbiome the status of "second brain," giving it salience also to Eastern medicine (Umeda 2019). While the precise communication pathways between the microbiome and the brain are not fully understood yet, they are believed to use neurological, immunological, endocrine (Ochoa-Repáraz and Kasper 2016) and vascular routes (Pennisi 2020). More specifically, it has been variably proposed that the microbiome's influence on the nervous system may be mediated through effects on immune homeostasis, immune response (Bray 2019) and inflammation (Pennisi 2020) via substances released by microbes into the gut, which then infiltrate blood vessels (ibid.); neurotransmitters in bacteria (Carpenter 2012); or through the most direct route between the gut and brain: the vagus nerve (ibid.).

The vagus nerve, which runs from the brainstem to the chest and abdomen, feeds information to the brain on the state of internal organs and can induce a more relaxed or heightened mood. This is why activation of the vagus nerve through deep abdominal breathing is frequently used to enable relaxation and heart rate stabilization. Many practitioners of Oriental Medicine will be conversant with this idea, as breathing is considered an important diagnostic and therapeutic instrument, which can be used to move stagnation, expel toxicity or pathogens, to nourish and more generally to balance energy across body systems. Interestingly, gut bacteria have been found to have the ability to influence activities in the vagus nerve (Breit *et al.* 2018), thereby affecting mood and behavior. This evidently adds another layer to interactions between the gut and brain and demonstrates the relevance of a holistic approach to the regulation of health axes that takes into account lifestyle factors and everyday choices, especially those we have the most ability to control. Such communication pathways illustrate also the multifaceted nature of the web of connections across body and mind, much like acupuncture's channels system.

ENVIRONMENTAL ISSUES

Pasteur's work on pathogens, while undeniably critical in both understanding and eradicating deadly microbes, inevitably placed much emphasis on destroying microbes—for good reason too, given the truly harmful effects of some, albeit a small number, of these (Yong 2017). However, since the time of Pasteur and arguably earlier in Chinese Medicine theory, a more nuanced interpretation of the impact of microbes existed, in which a person's "inner terrain" was seen to be a key determinant of the pathogen's ability to take hold (Rediger 2020). According to this proposition, the effect of disease-causing microbes is highly dependent on the environment into which they are introduced—a principle long part of Oriental Medicine both in the treatment of pestilent factors and in the prevention of ill health through strong Qi. Accordingly, the binary of "good" and "bad" bacteria is only partially accurate, given the complexity and dynamic nature of microbes. This is why the location in which a microbe is found is itself a determinant of what gives it the ability to have a positive or negative effect. Take for example a commensal bacterium such as *Staphylococcus epidermidis*, one of the many bacteria on our skin, with which in normal circumstances we happily co-habit. If the bacterium finds its way through intravenous medical devices, such as catheters, into the bloodstream of frail patients, for instance the elderly or premature infants, its impact can be catastrophic and lead to sepsis (Otto 2017), and if untreated death. Such opportunistic bacteria explain why hospital-acquired infections have become both common and difficult to deal with. The notorious *Clostridium difficile* bacteria often take hold following antibiotic treatments due to surgery or other hospital intervention, causing severe diarrhea and a rapid deterioration of a person's health, even though they are commonly found in healthy individuals. A key point here, therefore, is that outcomes are dictated by the type of bacterium, its location and the terrain, which in the case of patients on antibiotics has been stripped bare of a healthy microbial balance. These variables underline different microbial behavior patterns: "mutualists" that benefit both microbe and host, "commensals" that benefit the microbe without harming the host, "pathogens" where the microbe gains to the detriment of the host, or "pathobionts" defined as harmful in certain situations but beneficial in others (Fehervari 2019). This sophisticated understanding of the variability of bacteria and specific strains is not simply of academic interest; it has allowed therapeutic uses of bacteria to be developed, as exemplified by a particular strain of *E. coli* which is used in probiotic supplements (ibid.).

The idea that the "terrain" is itself crucial is, of course, intrinsic to Oriental Medicine. This is why in traditions such as *Five-Element* acupuncture, the practitioner uses multiple diagnostic techniques (Franglen 2014), spending first and foremost much time "enquiring" about patients' health and lifestyle in the form of a Traditional Diagnosis (TD) and using supplementary information from patients' physical manifestations, including color, sound, odor and emotion (CSOE) and pulses, in order to establish the root cause of their condition. The TD provides a thorough assessment of the patient's physical and mental landscape, covering aspects such as medical history, family health background, stamina, exercise, mental wellbeing, use of medication or supplements, diet, fluid intake and sleep. Each and every one of these can contribute to a person's terrain, which in the context of *Five-Element* theory would be seen as significant for an individual's *Elemental* imbalance. The environment created through everyday choices and circumstances therefore can be beneficial but can equally take its toll on a person. How taxing these choices and life events are on a person's health will depend on the baseline or *Elemental* constitution that is being worked from. Some people are lucky, with a genetic make-up that affords more leeway than others. This individual pre-birth asset, known as *Jing* in Chinese Medicine, is an inherited finite reserve of energy which is depleted at a rate that is congruent with lifestyle. The microbiome operates similarly as it is influenced by a combination of what we were born with and external factors.

Understanding the direct and indirect impact of internal and external environmental factors for the health of the microbiome throughout life is important since it shows us that individual choices can positively contribute to initiating changes necessary for the establishment and maintenance of good health. Dietary changes are rightly seen by many health experts as an area that is both within the reach of most and one which can have a considerable impact. Many books on the subject present compelling stories of a journey from disease to health, and how lives have been turned around through dietary changes (Campbell 2019; Collen 2015; Enders 2015) that include a wide range of pro- and prebiotic foods that fuel thriving bacterial communities. There is indeed a saying common to many cultures and therapeutic traditions that "food is the number one medicine" or that "we are what we eat," something that both Oriental Medicine and microbiome science help validate. For some time Western science has shown how food affects both mood and cognitive function. For example, the amino acids found in protein-rich foods are essential to the proper functioning of the central nervous system and are frequently used in sports supplements (Williams 2005) and alongside other slow-energy-release foods, such as oats and seeds, can help regulate blood sugar levels, thereby making us feel less anxious or irritable (Firth *et al.* 2020). What more recent microbiome studies are suggesting, however, is that our diet influences our microbiome, and this has the potential to significantly affect overall health, and critically too behavior and feelings (Bray 2019). Microbiome science has been able to show that host diet provides an avenue to changing the make-up and behavior of micro-organisms (Farrell 2019), and that in particular a high-fiber diet supports a greater diversity of bacteria in the gut (White 2019). Microbes thrive on such diets as they ferment the fiber, producing short-chain fatty acids (SCFAs) such as acetate, butyrate and propionate, which are believed to have wide-ranging health benefits (Morrison and Preston 2016; White 2019), such as lowering cholesterol and supporting heart health, fighting inflammation, protecting against cancer, regulating gut acidity, preventing a leaky gut, aiding brain functions, supporting a well-functioning immune system, and fighting against opportunistic bacteria and pathogens, while also helping regulate appetite and therefore a healthy weight.

Within and beyond nutrition, stress is another key environmental factor with disruptive effects on the microbiome and the inner terrain. Feelings of stress can be affected by, or interfere with, a person's diet, but stress can also take on many other forms and have many possible sources, including difficult relationships, work pressure, money worries, inadequate sleep, overexertion, chronic health issues or even side effects of medication. While disruptions to the microbiome linked to stress may be transient and of little consequence, over time they may lead to dysbiosis, which typically manifests as digestive problems, allergies, auto-immune conditions, fatigue or difficulty concentrating as well as providing opportunities for pathogens to take hold. The importance of stress in complicating the re-establishment of symbiosis cannot be stated enough. It is well known that stress has generally a detrimental effect on a person's health and may slow or reduce healing capabilities, increase susceptibility to disease, as well as encourage poor lifestyle choices, such as eating an unbalanced diet or drinking too much alcohol. Stress affects mood, behavior and cognitive abilities too and can manifest over time as anxiety, depression, anger, recklessness, poor judgment and decision-making, or memory problems. Each of these presentations can add to the burden of dysbiosis, creating a vicious circle with further patterns of ill health. Mental wellbeing is not just a manifestation of microbiome health, it can further influence it, which is why nourishing the balance means caring for our mental health too.

It is worth mentioning here the role played by sleep in patterns of microbiome disharmony. Poor sleep is often connected in some way to stress as either a cause or outcome, and when it comes to the gut microbiome, it has some very specific knock-on effects. Poor sleep is associated with changes in the levels of the hormones responsible for appetite regulation, with an increased level of the hunger hormone ghrelin and a decrease in the satiety hormone leptin (Kim, Jeong and Hong 2015). Eating more or too much and at certain times can all alter the balance of bacteria in the gut (ibid.), and this effect may be exacerbated by food choices when people are sleep deprived, notably foods with a high sugar content that are known to support the growth of bacteria associated with obesity (Mosley 2017). Poor sleep and changes to sleep patterns also affect circadian rhythms, and since there are indications that the gut microbiome itself is regulated through this natural inbuilt clock (Liang and FitzGerald 2017), imbalances in the microbiome may result not just from lack of sleep but also from irregular bedtimes and extended periods of wakefulness during the night. Although the reason for inadequate sleep may be beyond a person's control, for instance if they are working night shifts, have caring duties, or even noisy neighbors, it is important for practitioners and patients to be aware of the ramifications of sleep poverty and to understand the therapeutic value of seeking to resolve ongoing sleep issues whether through acupuncture or lifestyle changes.

DEFINING THE BALANCE

One of the challenges researchers, practitioners and patients face in the pursuit of a healthy microbiome that will deliver specific benefits to its host, is defining both quantitatively and qualitatively the required or even ideal microbiome profile for a particular individual. Even the effect of a certain balance of microbes may vary from person to person and at each site. When biomarkers do suggest correlations, these tend to be negative associations rather than positive, as homeostasis is possible even in the absence of so-called good bacteria (Sadanand 2019). It is not yet entirely clear either whether a disease triggers microbial variation, or the other way around.

More and more everyday food and healthcare products are flagging their "microbiome-boosting" qualities, whether body lotions, yoghurts, breads, chocolate or supplements, and while experts agree that there is a positive correlation between certain microbiome profiles and reduced likelihood of developing certain conditions, the microbiome—just like us—is very individual. This means that on the one hand, scientific and anecdotal evidence suggests patterns of health and disease associated with the microbiome's composition (Belizário and Faintuch 2018); on the other hand, precisely defining an individual's specific microbial requirements is often difficult. As such, our understanding and application of microbiome knowledge needs to strike a balance between objectivity, appreciation of individuality and awareness. Evidence to date indicates that microbiome dysbiosis can lead to specific disease progression, giving us at least some clues as to what appears to allow a harmonious relationship between host and microbiome. Firstly, a highly diverse community of microbes is considered important, as an overabundance of some strains of bacteria has been observed in patients suffering with both physical and mental health problems, including heart disease, Alzheimer's and obesity. Secondly, a stable and resistant microbiome with a rich genetic pool is considered beneficial to the balance of the microbiome and overall health (European Society for Neurogastroenterology and Motility 2021). In other words, even if the specific profile may not easily be defined, there is much that can be done to attempt to preserve and nourish a microbial balance and tap into the potential afforded by this innate system of medicine.

Trying to do this may seem futile to some, given how many unknowns there appear to be, while others may feel tempted to turn themselves into living experiments by throwing in every type of potentially beneficial probiotic. This is where the knowledge, wisdom, diagnostic patterns and qualitative assessments of health derived from Oriental Medicine provide an informative and practical approach to nourishing the microbiome. This can be seen as "fusion" medicine, an integrated approach where Eastern and Western ideas are brought together. The outcome measurement will be hard to argue with and is the same one that a classically trained acupuncturist uses within their treatment: asking patients how they feel and seeing improvements through the power of observation. The crucial word to remember amid this complexity is "balance," and the emphasis for any remedial work must be as such on reinstating equilibrium. Interestingly, in acupuncture, disease and ill health are considered a manifestation of an internal imbalance, and the role acupuncture and other modalities of Oriental Medicine may play in rebalancing microbial health represents an as yet little considered approach. Moreover, Oriental Medicine with its emphasis on dietary therapeutics and harmonious lifestyles (Deadman 2016) aligns well to core propositions within the microbiome science of the role of environmental factors, including our influence over these. Rather than something that we can simply throw all our tools at, the microbiome can be seen as a lens through which we can seek to understand and improve health outcomes, a framework within which we can learn to live more harmoniously with the world around us and ourselves, and an instrument we can all use to our benefit. Collectively, living well with microbes carries the potential to create a paradigm shift in our behavior and attitudes to health through self-care and preventative medicine. At an individual level, the microbiome can empower us to take charge of our destiny; it is our ultimate partner in health and disease: our microbiome is "us."

A FIELD OF NEW POSSIBILITIES

The universe of microbes is undeniably immense and the vast research in the field has helped establish taxonomic, functional, relational and methodological parameters that have shown how the microbiome represents a complex ecological system, a living community within the human body as well as a collective genome (Berg *et al.* 2020). The size and scope of influence of the microbiome, including interactions across body systems, has led some to refer to the microbiome as an "organ in its own right" (Amon and Sanderson 2017), or even "the forgotten organ" (Seo and Holtzman 2020). Such terminology is used to reflect the microbiome's involvement in, and regulation of, multiple and essential aspects of human health, including digestive, cognitive, behavioral, endocrine and immune functions, and its connections to other organs and the host's metabolism. Microbes are the very essence of life, and the significance and centrality of the microbiome justify for many its status as a vital organ. For some, however, a more precise interpretation of the microbiome is required, giving rise to terms such as "virtual organ" (Savage 2019), "superorganism" (Durack and Lynch 2019) or even "holobiont" (Simon *et al.* 2019). These perspectives underline well both the holistic and dynamic nature of the microbiome, its interconnections and the manner in which it is characterized by varying processes of colonization, resistance and cooperation.

The physical elusiveness of the microbiome will undoubtedly sound familiar to practitioners of Chinese Medicine, who may see some parallels here with *San Jiao*, the *Organ* "with a name but no form," which encompasses a collection of regulatory processes and homeostatic capabilities. Or perhaps they may even see in the microbiome the combined functions of the *Vital Substances—Blood, Qi* and *Jing*, or the changeable, complementary and opposing forces of *Yin* and *Yang*. The interrelationships that exist within and beyond the microbiome, and the allelopathic behavior of microbes—whereby certain micro-organisms suppress the growth of others—may, for a *Five-Element* acupuncturist, be seen to mimic the *Ke* and *Sheng* cycles, which allow balance and homeostasis to be maintained. While exploring these questions and assessing the relevance of Oriental Medicine for the wider understanding of the microbiome are the subjects of the next chapter, let us conclude here by remembering that what the microbiome and Oriental Medicine are both showing us is that there are always opportunities for change, and that health, rather than disease, is the norm we should focus on.

Chapter 2

Bringing the
Microbiome Home

As scientific research deepens our understanding of the microbiome and teaches us more and more about its powerful medicinal capabilities, so the imperative becomes stronger for practitioners of Oriental Medicine and their patients to find out how best to utilize this knowledge. Some readers may be wondering why these new insights matter, given that practitioners of Oriental Medicine already work within a well-established theoretical framework that delivers the exceptional results that are seen day after day in acupuncture clinics around the world. However, integrating microbiome science into the practice of Oriental Medicine is both pertinent and supportive of the therapeutic goals of acupuncture for several reasons.

Firstly, the microbiome, as we have already seen in the previous chapter, is hugely influenced by lifestyle choices and diet. Lifestyle recommendations have always played a valuable part in the practice of Oriental Medicine (see Deadman 2016). Acupuncturists are very aware that an acupuncture treatment can initiate necessary energetic changes in a patient, but that these will be enhanced and sustained over time by lifestyle choices that nourish and support the energy of each *Organ*, and that allow *Elemental* strengths to flourish. The recommendations provided by practitioners are much more than just "helpful advice." By bringing to patients' attention how to influence the course of their disease and the maintenance of their health and wellbeing, practitioners of Oriental Medicine are inviting patients to take ownership of their health, empowering them (Baker 2016), and helping them nurture the art of self-care and self-awareness for healing and longevity. This works in tandem with the acupuncture treatments themselves, which support, at body, mind and spirit levels, a patient's recognition of their ability to heal. It is through this transformation and renewed trust in the process of life that patients embark on a journey that reconnects them with the "self" and which helps them achieve a healthy existence, physically robust, emotionally balanced, and spiritually meaningful and fulfilling. The science of the microbiome, as we will see, fits very precisely within such perspectives.

Secondly, the microbiome offers practitioners of Oriental Medicine a golden opportunity to work with their patients in a language and with concepts with which many patients will be familiar or to which they will find it easy to relate. The microbiome and gut health in particular have become part of the everyday vernacular: from yoghurts "full of live bacteria" to body moisturizers that "support a healthy skin microbiome," many patients will have at the very least some vague notion that so-called friendly bacteria can make a positive contribution to our health, and many, if not most, patients will have noticed the growing presence of such products on shop and supermarket shelves. The microbiome provides a common language that facilitates communication between patient and practitioners, while offering a new perspective to patients' existing knowledge. Bringing Eastern and Western medical traditions together in this way also reflects the reality of the practice of Oriental Medicine in the West. Culturally and through mainstream care, many patients will have become deeply influenced by Western biomedicine and are often looking at Oriental Medicine as an alternative which fits into their referential framework, and that can be integrated into the conventional medical treatments they are receiving, or that can be used alongside other interventions.

Thirdly, from a diagnostic and treatment planning point of view, it is increasingly important for practitioners of Oriental Medicine not only to understand the energetic manifestation of dysbiosis and relate this back to the patient's medical history and lifestyle, but also to consider the significance of patients who are taking probiotic supplements. I am sure that I am not alone in finding that more and more patients are self-prescribing particular strains of probiotics with mixed outcomes. As the growing level of interest in the "microbiome" picks up pace, it is inevitable that this trend will continue. This additional dimension of the etiology of disease can accordingly help practitioners refine their diagnosis, treatment planning and food therapy advice.

Finally, understanding the microbiome through the lens of Oriental Medicine creates an interdisciplinary field that allows collaborative patient care and appropriate referrals to take place. Both Eastern and Western medicine share the common goal of helping patients, and by working together they create a highly personalized and truly holistic therapeutic approach. The aim of this chapter therefore is to provide meaningful ways for practitioners to integrate the two medical traditions, by considering theoretical and practical cross-overs between acupuncture and microbiome health.

FINDING THE MICROBIOME IN ACUPUNCTURE THEORY

The microbiome is not as such explicitly referred to in the ancient texts of Chinese Medicine that form the basis of acupuncture theory today. However, there are several ways in which it is conceptually present, thus underlining both the relevance and overlap between microbiome science and acupuncture theory and how it may be used to inform practitioners' diagnosis and treatment planning. To better understand how these systems intersect, let us examine the microbiome through specific theoretical tenets of health in Chinese

Medicine: *Yin–Yang*, *San Jiao* and *Mingmen Dantian*, *Jing*, external influences, specific disease initiation, *Elemental* imbalances and nutrition.

Yin–Yang

The different locations of the microbiome and the climatic variations these afford—including moist, dry, hot and cold conditions—are strongly reminiscent of the qualities of the mountains, valleys and marshes referred to in the name of different acupuncture points, providing valuable insights into how microbiome science can be integrated into an acupuncture diagnosis. The fact that every microbiome has its own ecology and that, for example, there will be differences in the make-up of the micro-organisms present on the arid landscape of the back of the hand compared with the crease of the elbow (Yong 2017) is particularly interesting from a Chinese Medicine point of view. More specifically, such contrasts reflect the energetic assessment and treatment used by acupuncturists, as some channels and acupuncture points are considered more *Yang*—palpable in the prone position and sitting on the exterior, more exposed, drier, and often darker skin areas—compared with the *Yin* channels, which are located on the anterior side with its usually softer, moister and paler skin.

The interdependent, complementary, opposing, mutually controlling and transformative forces of *Yin* and *Yang* are significant for the human microbiome, not just in reference to the contrasting body sites where communities of microbes are found, but also in relation to the mechanisms supporting the health of each microbial ecosystem. Each ecological community (Lozupone *et al.* 2012) represents a delicate balance that maintains homeostasis and can be seen to align to the *Yin–Yang* correspondences that exemplify the feedback mechanisms underpinning the mind–body continuum. In practice, this integration of *Yin–Yang* and the microbiome may help practitioners establish patterns of change, control and transformation, and may help explain even why micro-organisms behave the way they do in certain places. An example that springs to mind is the bacterium said to be responsible for acne (*Propionibacterium acnes*). This bacterium is a commensal which is part of the natural flora of the skin yet appears to trigger breakouts when suffocated in oily hair follicles (*New Scientist* 2016) as part of an immune response which creates a state of chronic inflammation (Collen 2015). The development of acne on the skin is most usually found on the upper parts of the body, and in particular the face or back. These areas can all be categorized as more *Yang* since they are on the skin, the *most exterior* point of defense against pathogens; on the head, the *highest point* of the body where *Yang* channels start (i.e. *Stomach, Bladder, Gallbladder*) or end (*Large Intestine, Small Intestine, Triple Heater/ San Jiao*); or in the case of the back, in a *posterior* location. Yet the bacterium *P. acnes* becomes problematic only in conditions linked to excessive oil production, increased testosterone levels (*Yang*) and when trapped in airless follicles that create an overly *Damp* and therefore extreme *Yin* environment. This leads to excess *Heat* (*Yang*), which the body attempts to control with more *Dampness* (more *Yin*). This interplay between excess *Yin* and excess *Yang* in the context of the skin microbiome aligns with the categorization used in acupuncture treatments for acne, which is energetically seen as "*Damp-Heat,*" an inflammatory state and mutually reinforcing cycle of *Dampness* and *Heat*, leading to chronic skin dysbiosis. It is of relevance to note additionally that, in Chinese Medicine, acne-related *Damp-Heat* imbalances are generally connected to *Stomach/Spleen* and *Lung/ Large Intestine* imbalances (Atkins 2013). This association is remarkably similar to findings from microbiome research that has linked gut health (which encompasses these *Organs*), immunity and acne (Lee, Byun and Kim 2019). For practitioners of Oriental Medicine, taking into account the role of the microbiome in health broadens the understanding of disease and may help elucidate important diagnostic details that may guide the practitioner's interpretation of the patient's history and experience of their condition.

Significantly, the theories of *Yin–Yang* and microbiome health are both underpinned by the principle of equilibrium, providing a common understanding that justifies the concurrent use of hands-on therapies such as *Five-Element* acupuncture, which seeks to re-establish internal balance by supporting the patient's *Element* and microbiome self-care, including harmonious nutrition, sleep and breathing. The Traditional Diagnosis undertaken at the patient's first appointment and the ongoing discussions that will take place between the patient and the practitioner provide opportunities to establish the physical or mental precursors of *Yin–Yang* and *Elemental* imbalances, and how these may suggest microbiome dysbiosis too, so that both the treatment and lifestyle recommendations can be calibrated to the patient's needs and healing. The focus of the practitioner's work remains as always on the patient rather than the condition and its labeling, yet this new knowledge is helpful since it may allow both seamless support for the patient and the resilience of microbial allies.

San Jiao and Mingmen Dantian

The concept of the microbiome, with its regulating and harmonizing functions, has a striking resemblance to the *Organ* in Chinese Medicine most directly associated with homeostatic functions, called *San Jiao (SJ)*, also known as *Triple Heater* or *Triple Burner*. For a long time, *SJ* was considered to have no direct equivalent in Western medicine, although more recently interesting parallels have been explored, notably by Keown (2014) who has likened it to fascia. The diagnostic and therapeutic significance of this *Organ* of the *Fire Element* has been the subject of debate for thousands of years, leading to different emphases in its use across various traditions of acupuncture. Some of the contemporary uses of *SJ* have focused on the channel's location. Starting at the nail base of the medial corner of the fourth digit and running posteriorly along the hand, arm, up to the shoulder and then on to the lateral surface of the neck to finish close to the temporal artery, it is today frequently used in the symptomatic treatment of hand, arm, neck and shoulder pain conditions as well as headaches. However, the names of the acupuncture points that can be found along the *SJ* channel tell us a much more interesting story about the channel and the *Organ* itself, symbolizing defensive functions, communication pathways and other spaces, notably gates (SJ-1, SJ-2, SJ-5, SJ-21) and foramen (SJ-15, SJ-22, SJ-23). Several points are specifically connected to water (SJ-4 "Yang Pond," SJ-9 "Four Rivers"), to temperature control (SJ-11 "Pure Cold Abyss," SJ-12 "Thawing the River") and to cross-generational interconnections (SJ-7 "Assembly of Ancestors," SJ-20 "Angle Grandson"). The acupuncture points therefore allude to both a source of energetic influence and a system of control and distribution that can be tapped into to restore the necessary balance and to ensure proper functioning of homeostatic functions and immunity. These are all aspects that are pertinent to the use and application of *SJ* in *Five-Element* acupuncture, where it is paired with *Pericardium* and which together with *Small Intestine* and *Heart* are associated with the most *Yang* of all *Elements*, *Fire*, yet are directly influenced by *Water*—the most *Yin* of all the *Elements*. Despite lacking a visible shape and formed instead of multiple relational operations, *San Jiao's* role explains why it is often likened to a thermostat that controls temperature, mood and a variety of regulatory and homeostatic functions, including endocrine, metabolic, digestive and immune. Interestingly, this description of *San Jiao* could equally be applied to the microbiome and may be seen as an early conceptual reference to the microbiome in Oriental Medicine.

The wider function attributed to *SJ* as the Official in charge of irrigation and the water passageways is especially relevant to the microbiome. This responsibility alludes to *SJ's* role in the digestion process and the distinct substances that reflect climatic and ecological variations on the head and torso. The top section—*Upper Jiao*—includes the head and chest, in which the concept of "mist" is used to describe the action of breathing, as the in-breath allows the lungs to expand and the blood vessels to absorb the oxygen that is taken to the heart, expelling de-oxygenated air on the out-breath. Directly below this section is the *Middle Jiao*, an area that extends to the navel. Here the analogy of a foam is used to represent the soupy substance made of partially digested foods and digestive juices found in the stomach and intestines, which is akin to chyme. The *Lower Jiao*, which covers broadly the pelvic area, is said to act like a swamp, where the clear and turbid are separated and waste removed. The combined *Three Jiao* accordingly define the different stages of digestion and key sites of microbial activity, including mouth, stomach, intestine and colon.

It is important to note also *SJ*'s connection to *Yuan Qi* (*Original Qi*), which is stored within *Mingmen Dantian*. This is the area below the umbilicus which stretches posteriorly to Du-4 "Gate of Vitality," described in some texts as the root of *SJ*, and anteriorly to Ren-4, an acupuncture point of extreme importance in Chinese Medicine as its no fewer than 24 different names testify (Willmont 2007). *Mingmen Dantian* is considered in acupuncture treatments and practices such as *Qigong* to be a highly critical energetic space and site of healing due to its connection with *Kidney* and the *Vital Substance* known as *Jing*, or *Essence*. The area around *Mingmen Dantian* highlights therefore its therapeutic and energetic significance in Oriental Medicine, which is nevertheless also closely connected to the microbiome, being similarly located to the highly rich concentration of influential micro-organisms of the intestinal microbiome. This overlap may suggest opportunities for practitioners in their treatment planning, and in particular the possibility that modulating *Yuan Qi* may support health in the microbiome and may even alter microbial activity.

Jing

The connection of *San Jiao* and the microbiome through *Mingmen Dantian* evidently brings into question how *Jing* relates to our internal microbial universe. *Jing* represents the transformative power which enables *Yuan Qi* to support vital functions directly and indirectly in the human body and is seen as a key determinant of health in Oriental Medicine, since it encapsulates the inherited potential for growth, reproduction and development. The substance produced by *Jing*, translated from Chinese as *Marrow*, enables health in the bones, spinal cord and brain. Depletion frequently manifests therefore as problems more commonly associated with the natural process of aging, such as weakening memory, vertigo, hair loss and graying, tinnitus, dizziness and brittle bones. It is often seen as a reflection of a person's constitution including their body's ability to correctly identify diseases and fight them off. As a result, a weakened *Jing* can manifest as a dysfunctional immune system which triggers allergies or autoimmune diseases. *Jing* naturally depletes with age and is especially vulnerable to undernourishment, stress and physical or mental overexertion. Its rate of attrition can, however, be slowed down through nourishing diets and appropriate lifestyle choices.

The determinants of *Jing* are remarkably similar to variables, including diet, stress, physical illness and mental health, that have been found to influence the microbiome (see Chapter 1). It is also interesting to see how overexertion, whether from working long hours, insufficient sleep or exercising too much—all of which are fairly common habits in Western societies dominated by a "work hard, play hard" culture—are explicitly connected to *Jing* depletion. This in itself matches up with the growing evidence that extreme and prolonged exercise regimens, such as those endured by athletes during training and competition, may dysregulate proper functioning of the gut microbiome (Clark and Mach 2016) because of the degradation of the mucus that supports microbial communities (Mosley 2017). This is not to say that exercise is harmful to the microbiome. Quite the opposite. Several studies have shown how well-adapted exercise and deep breathing do benefit microbiomes and how those partaking in regular suitable exercise have a different microbiome to their sedentary peers, and that this is linked to overall improved health outcomes (*The Scientist* 2019). This point is important to highlight as it shows that exercise is highly beneficial, but moderation is key and the level of activity should be relative to each individual. The moderate lifestyles which practitioners encourage their patients to pursue to support *Jing* are likely to positively influence the microbiome therefore, and arguably so will acupuncture treatments aimed at nourishing *Jing*. Furthermore, including microbiome-enhancing recommendations will help guide those patients whose *Jing* is depleting faster than expected, and could be targeted at patients showing signs of premature aging, including unexplained infertility, or to accelerate and sustain the pace of recovery.

Outside influences

Our ability to live in harmony with nature lies at the heart of *Taoism*, and as such susceptibility to ill health and changes in our mood at particular times of year or in response to certain weather conditions are used diagnostically and therapeutically in Oriental Medicine. In *Five-Element* acupuncture, *Horary* treatments are frequently called upon to harmonize an individual's response to energetic shifts, including climatic ones that occur throughout the day and year. In other traditions, such as TCM, climatic descriptors and influences are used to classify conditions that have precise qualities and origins, such as *Wind, Damp, Cold, Hot* and *Dry*. Climatic and seasonal influences on health and disease are not limited to Oriental Medicine, and the microbiome provides a striking demonstration of this. More specifically, the highly intelligent design of the microbiome has imparted evolutionary advantages to humans and animals that have helped them cope with seasonal changes, including times of food scarcity, such as during the winter months. This means that there appears to be a natural variation in microbial profile across seasons (Davenport *et al.* 2014), partly due to dietary changes and partly due to evolutionary survival processes. To achieve this, certain bacteria appear to extract more energy from the foods consumed, which means that their presence allows weight maintenance despite a significantly lower calorific intake (Yong 2017). In a situation where food is in abundance, an overdominance of such bacteria in the microbiome will lead to weight gain, even obesity. This helps explain why some individuals struggle to lose weight despite adhering to a restricted diet (Mosley 2017) and may also suggest a seasonal origin to dysbiosis. Added to this, the microbiome is readily affected by medication for seasonal allergies and other seasonal illnesses, exposure to pathogens such as winter flu, use of heating and cooling systems in homes, and changes to the amount of natural and artificial light we are exposed to on a daily basis.

Modern Western lives have distanced humans from winter food scarcity and natural seasonal flows. The problems this creates are compounded with the ritualistic overindulgences that accompany the winter festive period. Seasonal shifts and the associated microbial changes have combined therefore with cultural practices and social norms in a way that goes against natural rhythms, climatic and seasonal, and which may be contributing to patterns of dysbiosis with associated acute and chronic illnesses. This reflects a clear split from the *Taoist* principle that underlies acupuncture's therapeutic approach, whereby specific energetic shifts that are harmonious with nature and each season can reduce susceptibility to disease, especially when supported by lifestyles that include seasonal adaptations to diet, amount of sleep, exercise levels and time spent outdoors. Such variables are fully taken into account by practitioners of Oriental Medicine, yet the realization that these factors could have wider ramifications demonstrates the potential of interventions such as acupuncture to help compensate for microbiome imbalances that are rooted in seasonal or climatic factors.

Disease

Episodes of ill health, whether acute or chronic, can take a heavy toll on a person's health and wellbeing. In the practice of acupuncture, all too often we come across patients who have endured years of pain, digestive problems, fatigue and poor sleep, or who have more recently suffered an illness from which they have on the face of it "recovered" but who feel their vitality, stamina and general health are subpar. In acupuncture theory, episodes of ill health can deplete one or more directly affected *Organs* and importantly too energetic reserves, which weakens *Jing* and affects the speed and ability to recover and increase further susceptibility to illness.

Jing depletion and seasonal disharmonies are contributing factors to ill health, including in the microbiome, and are exacerbating factors in the harmful effects of pathogens and toxins, including viral infections, pollution, chemicals, additives or over-medication, which can further disrupt the microbiome for months at a time, perhaps even permanently. Pathogenic and toxin-induced imbalances in the microbiome are sometimes capable of spontaneously returning to a state of equilibrium in the case of a resilient microbiome, albeit perhaps in the form of an altered ecosystem (*New York Times* 2018). In some cases, however, there is persistence in the dysbiosis. A weakened constitution and an inability to expel fully lingering diseases are often to blame. While interventions such as probiotic supplements are often used in such cases, finding the right balance of microbial communities needed for each person is proving difficult (Eisenstein 2020b). This is why the procedure known as fecal transplant has shown tremendous results, since it reintroduces in its entirety a microbial community rather than a particular strand of bacteria (Yong 2017). Such treatments are still not readily available and may not be needed for every patient showing signs of chronic dysbiosis. It is valuable for practitioners and patients to consider then what role Oriental Medicine and acupuncture could play in restoring and maintaining a healthy balance in the microbiome. In particular, the core principle in both the theory and practice of *Five-Element* acupuncture, through its combination of prevention, maintenance and repair, provides an approach which addresses in a very targeted manner the precise needs of every individual, and by inference of their microbiome.

Elemental imbalance

In the *Five-Element* tradition of acupuncture, the diagnosis and treatment of a disharmony and its symptoms are centered around the principle that one of the *Five Elements*—*Water, Wood, Fire, Earth* and *Metal*—represents a dominant force in everyone, which governs an innate health in body, mind and spirit. The resultant classifications, known as *Wu-Xing* correspondences, provide a framework in which specific physical and emotional features (color, sound, odor and emotion), *Organs*, body parts, senses, times of day and year, directions of movement and even each stage of life relate to a specific *Element*. The *Elemental* typology is not intended to represent a finite reality, and practitioners of *Five-Element* acupuncture often find that some patients fit better within the model than others. The *Five-Element* approach therefore acts as a heuristic model for the observation of disease patterns which will guide an approach to treatment that seeks to re-establish balance by supporting an individual's dominant *Elemental* imbalance or "Causative Factor" (CF) (Worsley 1998).

The CF does not imply a rigid top-down model of classification, since all *Elements* should be recognized as present to a greater or lesser degree in everyone, manifesting at different times. The CF diagnosis allows the practitioner to focus on treating the main *Element* (and associated *Organs*) causing the imbalance to treat by proxy all other *Elements*, thus enabling each individual's potential to be realized. In other words, ill health is simply seen as a manifestation of a loss of harmony in a person which can be addressed by strengthening the patient's dominant *Elemental* energy, whilst situating this within their idiosyncratic life story. Two patients may present with the same condition but have a different CF, since it is their unique story combined with their *Elemental* make-up and the weaknesses that have developed therein that have led to the imbalances that need correcting. Interestingly, within *Five-Element* theory, the core diagnostic and therapeutic proposition is that a person's strengths and weaknesses are in fact one and the same thing, reflecting the duality of health and disease—an Achilles heel which nevertheless can reinstate harmonious functioning across all aspects of body, mind and spirit to become a "Guardian Element" (Franglen 2014). The human microbiome works in much the same way, as it can be seen as both the cause and solution to disease. As pointed out in Chapter 1, each person's microbiome is unique, shaped by genetic and environmental factors, much like a person's *Element*. The health benefits that microbial communities afford their host are dependent on ensuring the maintenance of a delicate balance of micro-organisms that are susceptible to disruption from external factors. If and when interruptions in the equilibrium are recognized and addressed appropriately, the potential to eradicate disease and maintain health becomes possible.

Not only can this analogy provide practitioners and patients of Eastern and Western medicine with a shared understanding and vision of the rationale and modalities of treatment, it also once again demonstrates the manner in which disciplines such as acupuncture can be integrated within microbiome therapies. We can only speculate at this point in time that *Five-Element* acupuncture's focus on correcting an *Elemental* imbalance may be able to positively influence the microbial balance too. While there is to date no research exploring this specifically in the context of *Five-Element* acupuncture, a number of animal studies (Jang *et al.* 2020; Qi *et al.* 2018; Wang *et al.* 2019; Xu *et al.* 2020) using interventions such as acupuncture, moxibustion or electroacupuncture suggest that these modalities are capable of modifying the microbial composition of the gut in diseased animals suffering from conditions including Parkinson's, obesity and ulcerative colitis, resulting in microbiomes similar to healthy animals and improving disease symptoms too.

Food, nutrition and diet

In Oriental Medicine, poor diet and malnutrition are designated as major causes of disease, and food is used as both curative and preventative medicine. There are many aspects that are taken into account in a Chinese Medicine diagnosis to explore with patients if, when and how their diet may be contributing to any ill health. To this end, practitioners of Oriental Medicine will fully assess the overall balance and individual suitability of a patient's diet since much emphasis is placed on both moderation and susceptibilities to specific conditions. Within this, aspects such as the temperature of the food consumed, the cooking method, the flavors emphasized in the diet, the overall energetic effects, the health of the *Spleen* (the *Organ* central to digestion in Chinese Medicine), the quantities consumed, the time of day of food intake, the regularity of meals, the person's emotional state when eating, the speed at which food is eaten and even the conditions in which food is consumed, are all seen as relevant. In other words, it is not just "what" we eat that matters, but also "how" we eat which allows the overall personal appropriateness of a diet to be assessed and where necessary adjusted.

While in Chinese Medicine every patient is diagnosed according to their individual pattern of excess and deficiency, whether *Yin*, *Yang*, *Blood*, *Dampness*, *Heat*, *Cold* or *Jing*, what transpires when looking at traditional diets still consumed today across East Asia is that food is rich in meaning. Food is often eaten in a way that emphasizes the notion of sharing, of showing love, of balance, harmony and nurturing oneself and others. With this comes a wide set of principles that are adhered to in the daily rituals of food preparation and eating itself, all of which ensure the digestive system is not overloaded and that food supports physical and mental health rather than hindering it. As a result, traditionally across East Asia, the tendency is for mealtimes to follow the rhythms of the day and seasons, with menus that draw on a variety of local, seasonal, freshly cooked and richly nutritional foods from the earth. Quantities are modest and commensurate with a person's level of activity. Such eating patterns contrast sharply with the modern Western diet designed for fast-paced twenty-first-century living where convenience, shelf-life, speed and functionality dominate. This split is reflected in health outcomes too. There is evidence suggesting that Oriental diets may explain the lower incidence, in several countries across East Asia, of certain diseases, including cancer, Alzheimer's and cardiovascular complications, as well as the higher occurrence of extended lifespans. Famously, in the islands of Okinawa in Japan, nicknamed the islands of "Eternal Youth," longevity has been partially attributed to a diet composed of a variety of plant-based foods eaten in small quantities (García and Miralles 2016). A varied diet with high levels of fiber from vegetables, fruits, nuts and seeds is now widely seen as the cornerstone of a balanced microbiome too, and with growing evidence that such a balance affects health outcomes positively (Makki *et al.* 2018), it is clear that the mantra in Oriental Medicine that food is medicine is as valid as ever.

It is important to reiterate that food therapy in Oriental Medicine goes beyond individual ingredients. This is a principle that must be considered when exploring the health benefits of nutrition on the microbiome as well as patterns of dysbiosis. For example, eating fast—a common feature in the rushed and pressurized environments associated with modern living—means that foods are not chewed properly, which not only influences the oral microbiome (Katagiri *et al.* 2019) and makes food harder to digest, it can also easily lead to overeating or eating more frequently. In Chinese Medicine, such overeating is associated with excessive *Dampness*, which creates a sluggish, heavy and sticky environment. It is easy to see then why periods of natural fasting, such as those that occur when meals are appropriately spaced out, enable the microbiome to return to a state of equilibrium (Paoli *et al.* 2019) as the *Dampness* clears and processes of spontaneous transformation can take place.

These aspects are discussed in greater depth in Chapter 4, but it is worth noting here the multifaceted nourishing potential of food and what this means for supplementation. The symbolism and cultural importance of food, together with the spiritual and emotional affinities of nutrition, as seen through the lens of Oriental Medicine, suggest something much deeper in how microbial communities and associated positive health outcomes can be supported. As noted in Chapter 1, microbiome medication is gaining momentum, notably in fields such as psychiatry where it is used to treat specific disorders, particularly depression and anxiety. While the benefits of such medication are not contested, for most people, especially for patients self-prescribing such supplements, knowing exactly what their individual needs are when it comes to their microbial profile is fraught with difficulty. The supplementary approach also perhaps misses other intangible positive influences on the microbiome that come from the cultural ritual of preparing food, taking part in eating a meal and the emotional effect this has, which together make eating nourishing in ways that go beyond functionality. In Chinese Medicine, when we enjoy eating nutritious foods with friends or family, or in a comfortable space, we are nourishing not just our body, but our mind and spirit too. Importantly, through the gut–brain axis, the positive effect of eating well in all manners arguably positively influences the microbes too. Food therefore offers us much more than just raw fuel.

CONCLUSION: A PERFECT UNION

Balance is everything when it comes to health, whether this is considered from the point of view of microbiome science or from a tradition within Oriental Medicine such as *Five-Element* acupuncture. Yet much like an *Elemental* balance, the microbial balance cannot be universally quantified or qualified, as it is unique and specific to every person. This provides a natural affinity and complementarity between the two systems, giving practitioners and patients new perspectives on ill health. Oriental Medicine has long recognized that illness is an opportunity for transformation, with endless possibilities for self-repair, while microbiome science is bringing to the fore the everyday opportunities for self-care that maintain good health and reduces the risk of disease. This convergence highlights a common principle of natural healing and homeostatic control that can be activated when needed. Restorative and maintenance instruments certainly exist within the microbiome, for instance in the form of autophagy, a process of repair through cellular destruction that occurs under energetic stress conditions, such as in periods of fasting or when exercising, allowing homeostatic functions and symbiotic relationships in the gut microbiome to be restored (Mosley 2017). Interestingly, this echoes somewhat the physiological stress that acupuncture needling mimics to trigger in-built healing mechanisms and to restore internal balance.

Such healing mechanisms underline the importance of energetic shifts in both systems of medicine and help us understand how the use of therapies such as acupuncture which manipulate, gather, strengthen and disperse *Qi* may be able to trigger the restoration of microbiome symbiosis. The manifold actions attributed to *Qi*, and the corrections that are sought—whether in terms of its speed, directionality, strength or overall quality—not only repair, restore and maintain health according to the principles of Oriental Medicine, it is conceivable that they can modulate the microbiome too. Using the two systems concurrently may allow the therapeutic effects from one to the other to be reinforced. This explains why it has more recently been suggested that the inflammatory effects on both gut and brain linked to both digestive and mental health provide unique therapeutic opportunities for acupuncture (Takeshita 2020) due to its anti-inflammatory effects. Acupuncture, it has certainly been suggested by recent animal studies (Jang *et al.* 2020; Wang *et al.* 2019; Xu *et al.* 2020), has the ability to directly modulate microbial communities. Although the precise mechanisms for this are not fully understood, it has long also been known that messages are conveyed between the gut and the brain, both upstream and downstream, and that the vagus nerve provides an important communication pathway between the gut and brain (see Chapter 3). Given the known effect on the vagus nerve of mind–body therapies, such as acupuncture and other practices including meditation, *Qigong* and *Tai Chi*, it is easy to see how many modalities within Oriental Medicine may be purposefully applied to address dysbiosis and maintain symbiosis. Of course, it goes without saying that given in-built feedback loops in the human body, the health of the microbiome can likely positively influence the outcomes of therapies used in Oriental Medicine, showing us once more the significance and impacts of the integration of these two systems.

Finally, it is essential to point out that this convergence of Eastern and Western medicine offers a strong reminder not only of our ability to heal, but of the personalized disease-prevention strategies at our disposal. The emergent knowledge in the field of the microbiome combined with the millennia of accumulated knowledge and practice in the field of Oriental Medicine provide a powerful set of tools for practitioners to help patients and for patients to help themselves. All too often the Western modern-day quick-fix approach to medicine focuses on illness and its suppression rather than the patient. This has led to patient expectations shifting away from personal responsibility and has resulted also in us often losing touch with our innate natural intelligence to repair and restore. The microbiome is putting Oriental Medicine's ideas of personal empowerment and the potentiality of healing and recovery center stage, providing scientists, doctors and patients alike with insights and know-how to break through barriers in medical and healthcare research, with an emphasis on personalized medicine that may be managed and enhanced by each individual.

Chapter 3

The Gut and Its Axes of Health

Over two thousand years ago, Hippocrates, the renowned physician of Ancient Greece who is considered the father of modern Western medicine and whose name is associated with the core professional principles and values set out in an oath used to ritualize medical doctors' entry into the profession, is said to have claimed that "All disease begins in the gut" (Lyon 2018). This idea has gained salience in recent years, as the pace and scope of scientific research in the field of the microbiome has highlighted the diverse and significant roles played by microbial cells within the gastro-intestinal tract in the maintenance of good health, and how microbial imbalances in the gut are linked to the development of conditions including inflammatory bowel disease (IBD), irritable bowel syndrome (IBS), diabetes, obesity, cancer, and cardiovascular and central nervous system disorders. While many factors can influence microbial balance, diet is certainly seen as key. This is because a person's diet has been found to alter the composition and behavior of microbial communities, meaning also that food choices have the potential to change the course of a disease (Farrell 2019).

The vast body of scientific work in the field has set in motion a thriving gut health industry, grassroots movements as well as expert and patient communities, characterized by a plethora of products, diets, supplements, websites, social media accounts and books seeking to explore, flag and fully exploit the potential of these new understandings of the gastro-intestinal (GI) tract's influence on our health and wellbeing. With gut-health theory and practice becoming so widespread, understanding its place within Oriental Medicine has never been so important for its practitioners and patients.

THE GI TRACT IN ORIENTAL MEDICINE

Hippocrates' proposition and the more recent discoveries about gut health resonate strongly with ideas relating to diet and digestion in Oriental Medicine, and to the all-important role of the *Earth Element*. The *Earth Element* and its associated *Organs* the *Spleen* (also referred to as *Spleen/Pancreas* due to its functions) and *Stomach* together encompass all aspects of food intake, the initiation of digestion (appetite, mouth, taste, saliva) and chemical breakdown of food, and thus also include key microbiomes such as in the oral cavity (Nelson-Dooley 2019) and stomach (Nardone and Compare 2015). *Earth* in *Five-Element* acupuncture emblematizes the nurturing mother–child relationship, connoting stability and nourishment. From this comes a circularity of movement and in particular the notion of "giving" and "taking," which is also reflected in the mixing and churning movement of digestion.

Physically, the *Earth Organs* lie in the middle area of the body. This location is symbolically reflected in the notion that the *Earth Element* is the center and the pivot around which everything revolves. This understanding means that the *Earth Organs* have significant influence on the health of peripheral areas too, assisting in the functions of *Blood* and *Qi*. Energetically, the movements and actions of the *Earth Element* encapsulate notions of "transportation," "transformation," "ripening" and "rotting." *Earth* symbolizes both the harvest time in late summer, with its golden, enrichening colors and soft tones, as well as points of transition across the year, such as the turn of seasons.

The breadth of *Earth's* influence is clearly described in key texts such as the *Yellow Emperor's Classic of Medicine*, which refers to the *Earth Organs* as an essential aspect of health during seasonal transitions, and also of the nourishment across all areas of nature, life and the human body (Ni 1995). As the *Earth Organs* are essential vectors of health, when disharmonies occur in them, wide-ranging mental and physical symptoms can appear, including excessive worrying, anxiety, lethargy, an inability to concentrate, muscular weakness, eating disorders or digestive problems.

The *Earth Organs* represent the first stage of the nutritive and digestive journey, and as other *Organs* become involved along the way, so do the energetic dynamics shift. In Chinese Medicine, the *Small Intestine*, which transforms food with the *Spleen* (Maciocia 1998), is associated with the *Fire Element* and its warming, uplifting, even purifying effect. Its task is to sort the pure from the impure both physically and mentally. As the *Small Intestine* is paired with the *Heart*, which carries out vital physical functions, including the circulation of blood and maintenance of oxygen levels, and emotionally enables an appropriate sense of joy, love and purpose in one's life, any depletion in the *Small Intestine* can have wide ramifications both physical and mental.

Further along the digestive journey is the *Large Intestine*, which receives food from the *Small Intestine* and which is perhaps best emblematized in acupuncture theory through one of the most widely used acupuncture points, LI-4 "The Great Eliminator." Partnered with the *Lung Organ* and part of the *Metal Element*, the *Large Intestine* has a descending energetic. The *Metal Element* represents the ability to take in and to let go, cleansing and ridding oneself of toxic or unwanted waste, both physically and mentally. It allows a sense of balance, clarity and focus. Emotionally, *Metal* has a poignancy to it, with a bittersweet quality much like the fragile and delicate beauty of the autumn which, as it progresses, leaves behind the joy and heat of the summer, taking us towards the stillness of winter.

Understanding the specific energetic of each of the *Organs* and *Elements* connected to the gastro-intestinal tract provides, as we will now see, an important basis for our exploration of the microbiome, as it offers a common platform to view the connections across the different sites of microbial activity and how these interact through axes that influence distinct aspects of health.

MAPPING THE CONNECTIONS

The gut microbiome's influence on other microbiomes, and health conditions more generally, underlines a number of axes of health (Schroeder and Bäckhed 2016). The best known in this nexus is the gut–brain axis (GBA), already briefly introduced in earlier chapters. In addition to the GBA, there is strong evidence of further communication pathways between the gut and other microbiomes, which provide symbiotic partnerships for the maintenance of good health. Interestingly, these networks closely resemble the balance that is maintained through the energetic cycles of nourishment (*Sheng*) and control (*Ke*) used in *Five-Element* acupuncture (see Franglen 2014; Worsley 1998), and which define specific relationships between individual *Elements* that interlink all aspects of health in body, mind and spirit. These relationships ensure good functioning across the entirety of a person. This also means that a disharmony within one *Element* can trigger further imbalances across one or more *Elements* and related *Organs*. Accordingly, in health, the *Sheng* relational paradigm allows the *Yang Organ* in one Element to nourish the energy of the next *Yang Organ* in the five phases diagram (*Metal* to *Water* to *Wood* to *Fire* to *Earth* to *Metal*), and similarly for the paired *Yin Organ* to do so for the next *Yin Organ* in the cycle. The *Ke* cycle for its part allows the *Yin Organ* to control or regulate the *Yin Organ* across the *Five Elemental* phases (*Metal* to *Wood* to *Earth* to *Water* to *Fire* to *Metal*). An insulting cycle is also sometimes considered (Maciocia 1998) whereby the affected *Organ* may trigger disharmony in the *Organ* from which it would normally be controlled.

These relationships are used by *Five-Element* acupuncturists to understand patterns of disharmonies linked to an individual's dominant *Element*, and together with other diagnostic tools, especially pulses, help define the source of any disruption along the body's energetic pathways and in one or more *Elements*. The schema informs a treatment strategy that goes beyond addressing presenting symptoms since these are seen as the end result of a disharmony.

Backtracking the source of the problem and even establishing the CF can be complicated as patients by choice or necessity may be masking the true state of their internal balance. There is culturally in many Western societies a tendency to "suppress" symptoms and depletion, often the result of a combination of bravado, shame or impatience as a "quick fix" approach combined with an expectation of "feel well, all the time" has been normalized. From an Oriental Medicine point of view, such suppression can allow a propagation of disharmonies. This is why, in *Five-Element* acupuncture, a specific protocol known as the "Aggressive Energy" (AE) drain is used at the very beginning of a course of treatment to remove conflicting signs and to reinstate balance. The protocol involves the shallow insertion with retention of acupuncture needles in the *Yin Back Shu* points which have correspondences to *Lung* (Bl-13), *Pericardium* (Bl-14), *Heart* (Bl-15), *Liver* (Bl-18), *Spleen* (Bl-20) and *Kidney* (Bl-23). As "negative energy" can only travel along the *Ke* cycle, the protocol, when performed correctly, allows any toxic energy to be cleared right back to its point of origin (see Franglen 2014; Jarrett 2006). This protocol allows practitioners to obtain a clearer picture of a person's health, and as such of the CF, before engaging in further treatments. It may also arguably be used diagnostically to look at patterns of dysbiosis across the different microbiomes, providing additional lines of enquiry for the practitioner and, therapeutically, opportunities for lifestyle recommendations, including dietary, that may be supportive of the patient's microbiomes.

There is, however, another treatment performed prior to an AE which is highly relevant to discussions of the microbiome. The *Internal Dragons* (ID) protocol, known also as *Seven Dragons*, is frequently used within *Five-Element* acupuncture in cases where there is strong evidence of a dysfunctional relationship between body and mind, often arising in patients who have experienced trauma, and manifesting in an inability to control thoughts and actions, with for example obsessive behavioral patterns, extreme anxiety or panic attacks. The trauma may be linked to a particular event, severe illness, medication, use of recreational drugs or physical depletion, including from highly restrictive and nutritionally poor diets. There may also be milder cases where the cause is not evident, but where patients experience a sense of detachment from their "self," unable to connect with the experiential quality of their life or with others around them. Practitioners may simply notice signs of incongruity in these patients with a mismatch between their thoughts, what they say and their actions. There may also be an emptiness or avoidance when looking a patient in the eye. Interestingly, I have noted how digestive health issues or problems pertaining to appetite regulation and food intake are common among patients requiring the use of the ID treatment. The acupuncture points used in the ID protocol are mostly along the *Stomach* channel (St-25, St-32, St-41). The protocol includes also a master point located just below Ren-15 which can be found in the epigastric region of the abdomen. This area is close to the dynamic environment of the gastric microbiome, and to key digestive organs modulated by the gut microbiome and influenced by diet, stress, drugs and diseases.

The need for this protocol is far more common than may be assumed, and the patient's presentation is key to determining its use rather than relying on a history of trauma. The positive changes that are initiated through this treatment are remarkable and many patients, some of whom may have tried other therapies, sometimes even for years, have themselves been surprised at the difference this treatment and the follow-up AE protocol have made to their health and wellbeing. Time after time, practitioners note the visible changes in their patients following the treatment, including their demeanor and the color or tone on their face. Physical symptoms, notably gastric health, are often greatly improved from these two treatments alone. Emotionally and spiritually, patients' feedback is simply expressed as feeling at home with oneself. It is almost as if the child-like quality of being connected to the present, an innocent joy spirited by a belief in the infinite possibilities and the quasi-magical unfolding of life that affords an effortless ability to move through what are normal emotional states, becomes possible once again.

The application of protocols such as these demonstrates well both the connection between body and mind, and the virtuous circle they form. Body and mind are in constant communication, with information being sent via nerves out of the brain to different body parts, as well as from the body into the brain, creating a mind–body feedback loop. What is emerging from microbiome research is that micro-organisms are very much part of this cross-organ communication network. While this is only the subject of speculation, it may be possible to contend that because the points used in the *Seven Dragons* protocol are connected with the *Stomach*, which in the context of Western medicine is also considered a key organ of the digestive tract and line of defense against harmful pathogens, and in the case of AE to the six interconnected *Yin Organs* that ensure homeostasis, the improvements noticed by patients are also linked to microbial readjustments that are capable of re-establishing symbiotic communications for the maintenance of health across body and mind.

Such questions can only be fully answered by further research. However, the way connections between body systems are enabled through the existence and activities of micro-organisms does suggest at least some degree of alignment to the relational health frameworks used by acupuncturists. This intersection between inter-microbiome communication and Oriental Medicine is especially intriguing when looking more closely at two particular axes, namely gut–brain and gut–lung.

THE GUT–BRAIN AXIS

Micro-organisms, predominantly bacteria, are present across many sites of the human body, although it is the GI tract that has the biggest share, with an impressive diversity of microbial communities present along the tract from the esophagus right through to the colon (Dekaboruah *et al.* 2020). Profiling of the microbiota's composition and activities across the organs forming the GI tract suggests a complex set of gut–brain interactions that support homeostatic functions (Carabotti *et al.* 2015). As a result, the gut microbiome is also increasingly seen as a key player in driving brain disorders (Willyard 2021) such as autism, Alzheimer's, anxiety, Parkinson's and depression. Given the significant impact on quality of life for sufferers and their families and the growing prevalence of such conditions, understanding gut–brain interactions is more than mere scientific endeavor, since they may provide vital clues for new therapeutic interventions. Interestingly, despite its more recent revival through microbiome research, the notion of a gut–brain axis was in fact central to medical thought throughout the eighteenth and nineteenth centuries in places like Britain, when doctors considered the stomach to be closely entwined with a person's emotional health and to be possibly even the root of all bodily and mental disorders (Miller 2018). Although such views had for some time lost favor in medical circles as an arbitrary mind–body divide came to dominate Western medical practice, the proposition that the composition of the gut microbiome influences brain function and mental health marks a much-needed shift away from a tendency to consider matters of the mind and body separately. Importantly too, this shift brings with it a recognition of the close interconnections that exist between neurological functions, behavioral patterns and physical presentations and thereby is helping validate the use of interventions such as acupuncture, while allowing patients to reframe their understanding of the relationship between body and mind through the rationale of Western science.

The idea of a connected whole in body and mind is integral to Oriental Medicine, since in this system all *Organs* perform both physical and mental tasks. This partly explains why functions of the brain are not reduced to a single physical location or organ, even if specific mechanisms exist to support it. *Jing*, or *Essence* partly inherited, partly sustained through sound lifestyle choices, discussed earlier, reflects one aspect of this multiplicity as it feeds the brain, thus playing a vital role in mental health and brain function. However, activities that engage functions of the brain are connected to *Organs* found elsewhere in the body, as expressions of the mind through emotions and virtues are linked to the five *Yin Organs*. Anger and benevolence pertain to the *Liver*, joy and compassion to the *Heart*, sympathy and sincerity to the *Spleen*, grief and justice to the *Lung*, and fear and wisdom to the *Kidney*.

The mind itself is connected to *Blood* in Chinese Medicine, and the common condition associated with its impairment (*Blood Deficiency*) manifests in wide-ranging mental dysfunctions which can vary in severity from mild to more severe, and include anxiety, general nervousness, low mood, dissatisfaction, irritability and lack of self-worth. The link between mind and *Blood* finds correspondences also in the gut–brain axis, through Chinese Medicine's understanding of how *Blood* is also highly dependent on good nutrition and adequate rest, aspects that are known to affect the composition of the microbiome too. Significantly, in Chinese Medicine, the onset and existence of *Blood Deficiency*, which is readily diagnosed by acupuncturists through thin and often tight pulse qualities, a pale tongue and physical symptoms such as pallor, dizziness, blurry vision, dry skin or hair, tight muscles, menstrual problems and mental wellbeing problems, may be caused by parasites, bacteria and viruses which interfere with the immune system and inhibit proper *Blood* production (Leggett 2014a, 2014b). Oriental Medicine's conceptualization of the influence of *Blood* on both physical and mental health offers an interesting perspective on the associative functions of the gut and brain, and how both diet and specific bacterial, parasitic and viral disruptors may be involved in patterns of dysbiosis. Moreover, several recent studies have suggested that blood may not be a sterile environment (Castillo *et al.* 2019) and that there may be a "blood microbiome" susceptible to disruption and capable of initiating disease. For practitioners of Oriental Medicine, a treatment strategy that supports the health of *Blood* may be important therefore to nourish, moisten, repair, cleanse and stabilize not only body and mind, but microbial sites too.

It is crucial to note that in addition to the emotional qualities associated with each *Organ* and the connections between *Blood* and mind, it is the *Heart* or *Shen* which is considered the residence of the mind in Chinese Medicine, encapsulating both mental (emotional) and, importantly, spiritual aspects of human beings. As is the case with the gut–brain axis, *Heart* and the gastro-intestinal tract are connected in several ways. The first is through its *Elementally* paired *Organ*, *Small Intestine*, and the second is through the *Sheng* generating cycle, as *Heart* nourishes *Spleen* (which includes the functions of the pancreas in Chinese Medicine).

In the context of a microbiome mapping, the generating *Sheng* cycle could suggest that health within the *Heart*, and therefore mind, is indispensable for the *Spleen* and therefore the digestive system as a whole. It is also the case that the transformation of *Food Qi* into *Blood* is said to take place in the *Heart*, and as such demonstrates a close relationship between food, nourishment, mind, spirit and *Blood*. Conceptually, a similar link exists within Western medicine through a gut microbiome and cardiovascular health correspondence, as the presence of particular genes in the microbiome appears to correlate with lower blood cholesterols, while a diet rich in fiber on which the bacteria feed increases the formation of short-chain fatty acids (SCFAs) that contribute to the regulation of blood pressure (Harvard T.H. Chan School of Public Health 2021a).

Such interdependencies demonstrate not only the role played by food and nutrition, but also how nourishing directly or indirectly *Heart/Shen* through acupuncture treatments may be supportive of harmonious gut–brain communication and may enable better functioning and balance within the gastro-intestinal microbiomes. Introducing microbiome science in this way into an acupuncture treatment not only allows practitioners to make use of a multidimensional understanding of the functions of the brain and mind as established through the theory of Chinese Medicine, it allows them to apply their knowledge and observations throughout the treatments in a manner that is consistent with the patient's own experience of disease or ill health. This approach further enlightens the often complex patterns of imbalance of each individual patient, offering a precise and intelligent understanding of symptoms and their treatment, while harnessing the potential for healing of the bidirectional communication between the gut and the brain. There is another strong rationale for paying particular attention to *Heart* and *Blood* within a strategy that seeks to enhance the potentiality afforded by the strength of the microbiome. *Heart* (also called "The Emperor" or "Sovereign Ruler") and *Blood* are considered alongside *Jing* powerful determinants of a person's constitution. The *Heart* significantly works synergistically with the *Kidney*, emblematizing the balance of *Fire* and *Water*, *Yang* and *Yin*, and all fundamental aspects of a person's existence, through the delicate balance that is played out between "self" and "destiny."

The gut–brain axis may therefore be seen from an Oriental Medicine point of view as more than a relational pathway of two physical entities, since other *Organs* and aspects connected to the mind, especially *Blood* and *Heart*, are critical too. The GBA represents in many ways the foundational basis from which good health—physical, mental and spiritual—can emerge.

THE GUT–LUNG AXIS

Over the last few decades, new investigative techniques have been providing an increasingly fine-grained understanding of the human microbiomes, including in the lungs. This work has helped ascertain both the existence and importance of the lung microbiome, and significantly too its connection to other microbiomes, notably the gastro-intestinal microbiome forming what is known as the gut–lung axis (GLA). The GLA is instrumental in human health as it is characterized by direct and indirect communication pathways between the gut and lung. These are closely involved in immune responses, meaning that the composition of the gut microbiome itself can influence the progression of respiratory diseases (Enaud *et al.* 2020). Studies of germ-free mice (Lo *et al.* 2021), which showed a greater susceptibility to disease among this specially bred group of animals, had already suggested a link between immune responses and the gut microbiome. What the intersection of gut health and lung-related diseases is now underlining, however, are the intricate ways in which the gut microbiome may have the ability to shape disease outcomes. For instance, studies of mice suggest that the use of antibiotics, known to significantly reduce microbial diversity in the gut, may reduce immune responses to flu (Enaud *et al.* 2020). The gut–lung axis has also been explored in the context of Sars-Cov-2, where it has been found that patterns of gut dysbiosis seem to be involved in the disease, suggesting respiratory viral infections affect and are affected by gut homeostasis (Sencio, Machado and Trottein 2021). It is worth noting that even the way we breathe can affect the health of the microbiome, as breathing excessively through the mouth can introduce imbalances in the oral microbiome which in turn can disrupt microbial balance in the gut (Nelson-Dooley 2019).

Importantly, as with the GBA, the GLA communication is bidirectional. For instance, viral illnesses, like influenza, and some inherited lung conditions, such as cystic fibrosis, appear to lead to changes in the gastro-intestinal microbiome composition and correlate with specific bacterial profiles in the gut, while dietary modifications for their part have been found to alter the composition of the lung microbiota (Enaud *et al.* 2020). Interestingly too, the predominance of certain bacteria in the gut is also found to be reflected in the microbial profiles of the lungs. The GLA hence helps explain why a lack of diversity in the micro-organisms present in the gut of newborn babies can lead to the development of childhood asthma (Milani *et al.* 2017) or how the overgrowth of certain gut bacteria may be associated with the severity and progression of various lung conditions. The relationship between the gut and the lungs is more than a connection, however, and should instead be seen as a partnership where one influences the other, making each one susceptible to the effects of disruption in the microbiome of the other.

In Oriental Medicine, there are a number of ways in which we may see such interconnections between the gastro-intestinal tract and the lungs, and the important role these play in proper immune function in particular. The *Lung* in Chinese Medicine is classified as the *Yin Organ* of the *Metal Element* and is paired with the *Large Intestine*. While each *Organ* has its own specific functions, these paired *Organs* work closely together. For example, it is the *Qi* from the *Lung* that encourages the descending action needed for excreting waste, meaning that constipation is sometimes considered a manifestation of *Lung* weakness. Significantly too, irregular bowel movements are understood to impact on a person's breathing and can lead to shortness of breath, as *Lung* function is impacted by stagnation in the colon. In the *Sheng* cycle of *Five-Element* theory, the digestive *Organs* of *Stomach* and *Spleen/Pancreas* directly influence the health of the *Lung/Large Intestine* pair. As such, a deficient patient with a *Metal CF* diagnosis often greatly benefits simply from tonification points (Lu-9, LI-11), or by ensuring appropriate strength in the *Earth Element* through horary acupuncture points (Sp-3, St-36) as well as dietary recommendations.

Lung function is of course seen as essential to support the proper functioning of all organs/*Organs* and body systems in both Eastern and Western medicine, allowing for the inhalation of clean air, and removal of waste product in the form of carbon dioxide. It is through deep breathing that more air enters the lungs so that blood vessels can better absorb the oxygen that is taken to the heart, slowing down heart rate and stabilizing blood pressure. This explains why in the meditative and deep breathing practices commonly used within Oriental Medicine—including *Yoga*, *Tai Qi* and *Qigong*—so much importance is placed on the power of the breath. These practices allow a seamless connection between the movement of breath, mind and body, moving both *Qi* and *Blood*, dispersing any stagnation, nourishing all *Organs*, and thereby enabling physical and mental repair and the maintenance of good health.

Qi provides useful insights into how lung function influences other body systems and their microbiomes, including in the gut. *Qi*, an immaterial and multifunction substance central to Oriental Medicine, manifests in various forms and energetic forces, and works similarly to microbiome communication pathways. *Qi* can transform into both defensive (*Wei Qi*) and nutritive forces (*Ying Qi*), through a gathering (*Zong Qi*) process that connects the energy derived from food (*Gu Qi*) and the air breathed through the work of the *Lung*. Importantly, *Zong Qi* is aided by the *Original Qi* (*Yuan Qi*) stored in *Mingmen* (as discussed in Chapter 2) to create the *True Qi* (*Zhen Qi*) that circulates through the channels. As the strength and quality of a person's *Qi* is assessed through pulse diagnosis, some interesting questions arise on the diagnostic and therapeutic possibilities this offers acupuncturists: to what extent is the health of the microbiome detectable through pulse diagnosis and other signs of *Qi* deficiency, such as shortness of breath, sweating, pallor and tiredness? Can the gut microbiome be improved through treatments that seek to ensure the vitality simultaneously of the *Lung* and of the *Organs* of the gastro-intestinal tract?

Although these questions are yet to be answered, they certainly offer a different perspective on the progression and treatment of disease, and on the impact of the acupuncturist's work in re-establishing balance and homeostatic function in the microbiome. From experience of treating patients who suffered from symptoms resulting from gut dysbiosis or whose presentation was suggestive of it, I would argue that there are often indications of specific pulse patterns and interestingly of *Lung Qi* deficiency in particular. As such, harmonizing pulses, providing lifestyle recommendations that include mindful deep breathing, spending sufficient time exposed to fresh, clean air and minimizing pollutants, whether perfumes or toxic substances, and seeking more generally to improve the strength of *Lung Qi* may prove a particularly useful way to enhance microbial diversity across the gut and lung microbiomes. This therapeutic approach may also be justified by suggestions that ingested air pollution and toxins may be a contributing factor to changes in the gut microbiome and to the development of inflammatory bowel disease (Salim, Kaplan and Madsen 2014).

It is worth pointing out also that the *Lung* in Chinese Medicine is said to control the skin. The process by which this function is enabled is especially relevant when considering the interrelationships between the microbiome of the lungs and the gastro-intestinal tract, as *Lung* energy, in acupuncture theory, is understood to distribute fluids that have been received from the *Spleen* all the way to the skin. In health, the skin has a balanced level of moisture that ensures that, as the most external organ, it can perform well its protective role. In cases of gut dysbiosis, it is common for patients to experience chronic skin conditions, notably eczema and psoriasis, while dietary and lifestyle changes that target the health of the microbiome appear to provide a marked improvement in skin health (Campbell 2019). The profile of the microbiome present on the skin (Byrd, Belkaid and Segre 2018) is, however, also dependent on direct exposure and contact to micro-organisms. The skin microbiome is frequently disturbed by the everyday detergents and chemicals that disrupt a healthy flora. This suggests the need for a wider understanding of food therapy, which in Oriental Medicine, as already pointed out in Chapter 2, includes all aspects of eating and digesting. Reinstating diversity and balance may need to come therefore not only from food choices themselves, but also from food preparation and cooking practices that encourage exposure to a diversity of micro-organisms, such as when washing the soil off fresh vegetables or preparing a homemade sourdough made from wild yeast. Such practices, as will be discussed again later on, may directly influence the health of the microbiome, and positively impact on mental health too, thereby further enhancing thriving microbial communities.

CONCLUSION: THE PARTS THAT MAKE THE WHOLE

The GBA and GLA go some way to showing how the interconnectedness of body systems found in Oriental Medicine is both relevant and useful in understanding mechanisms of gut dysbiosis and their effect on overall health. It is worth noting, however, that these are just some of the links across body microbiomes. As other microbiomes become better understood, further axes may gain salience for practitioners of Oriental Medicine or other healthcare professionals seeking to integrate Eastern and Western medical perspectives. For instance, the progression of liver diseases has more recently come to be seen as linked to gut dysbiosis (Fukui 2019), which may be linked to the *Ke* controlling or "insulting" axis between *Lung*, *Liver* and *Spleen*.

These interconnections provide a detailed understanding of feedback loops between body and mind that have long been understood, notably the "fight or flight" mode—associated with fear and panic, racing heart and an urgency to open the bowels, and the "rest and digest" mode, which creates a sense of peace and relaxation. Importantly too, this understanding of a highly and intricately connected set of "components" that make the "whole" (Kaptchuk 2000) offers a key lesson on accessing the full potential afforded by having a harmonious and thriving ecosystem. It demonstrates that the way to nourish the balance of the microbiome is much broader and more holistic than may be assumed, reflecting fundamental principles of health in Oriental Medicine: breathing, eating and sleeping well, as well as having positive and meaningful relationships with ourselves and others.

Chapter 4

Food Energetics and the Microbiome

Microbiome-friendly nutrition, with its focus on fresh, natural and plant-based foods that are rich in fiber and vitamins, is seen as playing a vital role in achieving and maintaining good health. This emphasis has much in common with Oriental Medicine's nutritional approach, in which vegetables, fruits and grains are prioritized according to key energetic principles. This complementarity makes Oriental Medicine ideally placed to enhance the application of microbiome dietary science. More specifically, as this chapter explains, the *Elemental* associations and the concept of "food energetics" used in Chinese Medicine create a holistic dietary framework that can be used to both address specific patterns of dysbiosis and help practitioners and patients develop a highly individualized approach to nourishing the microbial balance.

THE ENERGETICS OF FOODS

Foods in Oriental Medicine are classified according to their heating, cooling, drying or moistening "energetic" effect on both body and mind. Each of these influences is connected to seasonal attributes, times of day and climatic variability, and when used appropriately can both benefit specific *Organs* and address patterns of energetic excess or deficiency (see Leggett 2014a, 2014b; Pitchford 2002). Above all, "food energetics" is a therapeutic system that takes into account simultaneously internal dynamics and individual patterns of health and disease, as well as the external environment, to ensure harmonious living in the deepest and broadest possible sense.

The food energetics model provides a sound and logical approach to re-establish and maintain microbiome symbiosis, since its goal is to ensure homeostatic functions across body systems and in the *Organs* connected to the GI tract. It is a framework which also helps unravel how dysbiosis in the microbiome occurs and how it may be tackled. This chapter explores, therefore, the principles of food energetics to demonstrate both how they pertain to the microbiome and how they may inform gut health eating practices. To do this, it is important to examine first those imbalances and other diagnostic and therapeutic considerations linked to diet and lifestyle in Chinese Medicine, and which have a particular significance for gut health, starting with *Dampness*.

DAMPNESS

In Oriental Medicine, the *Earth Organs* are considered pivotal to the digestive system, and of particular relevance to patterns of dysbiosis is a condition closely linked to these *Organs*, known as *Dampness*. The condition is a *Yin* disorder and is a fairly prevalent presentation among patients seen in acupuncture clinics. It is usually associated with the excessive consumption of *Damp*-forming foods, such as saturated fats, pastries and biscuits, some breads, sweets, sugar and artificial sweeteners, even fruits that grow in very hot climates and juices made from these, beer, as well as rich and fatty meats. Such foods are regularly and often excessively consumed in Western diets, even on a daily basis. *Dampness* is more generally exacerbated by overeating, sedentary lifestyles, or frequent consumption of chilled, raw or cold foods.

Dampness can manifest in a variety of stagnant energetic presentations including weight gain, cysts, tumors, a feeling of heaviness, puffiness, joint pain, poor digestion, fatigue or low energy, as fluids and mucus build up, accumulate and become fixed, resulting in congestion, stagnation and overgrowth, with swamp-like conditions. *Dampness* can show at a mental level in the form of brain fog, excessive worrying, poor concentration, memory problems and a sense of heaviness in the head, often accompanied by headaches. Diagnostically, acupuncturists may note such tell-tale signs and symptoms, but it is also picked up through examination of a patient's tongue and pulses. There is in some cases a distinctive energetic movement in patients and also an odor that can be further suggestive of *Dampness*.

While there can be individual predispositions or lifestyle triggers, *Dampness* can result from ongoing gastro-intestinal conditions, as well as infections that linger (Stephenson 2011), and is likely to be present in presentations such as post-viral chronic fatigue syndrome (CFS), myalgic encephalomyelitis (ME) or even long Covid. It can even arise from living in a damp environment. Some body areas and health conditions, such as pregnancy, which are by nature already *Damp*, are particularly susceptible to excessive *Dampness*, which is why poor digestion, bloating, nausea, build-up of fluids or vaginal infections linked to an overgrowth of yeast are common during pregnancy.

Dampness can be either cold or hot in nature, and the contrast can be compared to the experience of a hot and humid summer's day compared with cold, foggy conditions, such as during the winter or when a thick blanket of fog rolls off the sea into coastal areas. Despite their temperature difference, the main characteristics at both ends of this hot and cold spectrum remain the same: a slow, stagnant, clogging, swelling and heavy energetic movement. These features give the condition the ability to persist and worsen over time, often creating a mutually reinforcing cycle of disharmony between *Dampness* and *Damp*-forming lifestyles and habits. Due to the significant disharmonies and ill health linked to *Dampness*, practitioners of Oriental Medicine will often seek to resolve *Dampness* directly through their treatment strategy and by encouraging patients to both limit the intake of exacerbating foods and use countermeasures to reduce or resolve any excess of *Damp*.

Although most closely associated with the digestive system, *Dampness* can be present in other *Organs* in Oriental Medicine, notably *Heart* and *Lung*. Here the patterns of diseases remain the same: a stickiness or heavy mucus that clogs the system, preventing *Organs* from working properly. In the *Heart Organ*, a disturbing illustration of this comes from the changes that may be observed over time in a person suffering with Alzheimer's disease, as a heartfelt barrier to communication develops in the patient. It is said in Chinese Medicine that the health of the *Shen* (Heart-Mind) can be gaged by looking a person in the eye, and that Alzheimer's is characterized by "*a mist on the Heart.*" Anyone who has looked in the eye of a friend, loved one or patient suffering with Alzheimer's will have noticed the absence of that inner light and sparkle that reflects the *Shen* or inner *Spirit* which, by contrast, is readily found in the eyes of a child. The mist or fog on the *Heart* inevitably also disrupts the *Water/Kidney* (*Yin*) and *Fire/Heart* (*Yang*) balance and explains both this inability to communicate appropriately and the mental confusion that accompanies the disease, triggering unrest, fear and panic in sufferers which often peaks in the late afternoon during the *Kidney's* peak hours.

Dampness connected to *Lung* will manifest as nasal discharge and accumulations of mucus in various forms. Through *Lung's* connection with the skin, it may also trigger yeast or fungal skin infections. As the body sometimes tries to rid itself of the excess of moisture, its attempt at overriding the *Dampness* generates *Heat*, which leads to even more *Dampness* being produced, and more counter-*Heat*, resulting in itchy, dry, sometimes shiny or combined presentations in the form of, for example, fungal skin conditions, allergic rashes, herpes or eczema.

Dampness is a condition that is especially relevant to any discussion on supporting everyday health, partly due to the prevalence of *Damp*-related health problems seen in acupuncture clinics and other medical settings in many countries around the world, and partly because there is a clear overlap between conditions typically associated with gut dysbiosis and *Dampness*.

DIAGNOSING DAMPNESS FOR MICROBIOME HEALTH

The similarity between *Dampness* and conditions associated with microbiome dysbiosis, coupled with the important role of Chinese Medicine's food energetics in the management of *Damp* conditions, provides a powerful argument for practitioners and patients to consider both signs and possible triggers of *Dampness* as part of any strategy to reinstate symbiosis and to maintain a healthy microbiome balance.

As *Dampness* can take different forms—hot, cold and congealed—its diagnosis and treatment will involve assessing fully the patterns of disharmony. This is important to determine the root cause of the *Dampness* and contextual appropriateness. Many clinicians will have seen patients who, as they "heat up" due to heightened stress or demands from work, find *Damp*-forming foods to be irresistible, even soothing. This is because *Dampness* has a slowing-down effect which provides some sense of immediate and temporary relief in situations of stress. In the long run, however, disproportionate consumption of *Damp*-forming foods contributes to further disharmonies, including more *Heat* and *Damp* conditions. A similar principle occurs in the communication of neurons in the gut microbiome, whereby an overgrowth of particular types of micro-organisms can be the very reason a person struggles to resist sweet cravings. Reaching out for that sugar fix is not just habitual behavior, it can also be explained through the microbes' attempts to establish their dominance (Mosley 2017), sending messages to the brain to eat sugary foods that give a temporary spike in energy and surge of dopamine.

It is essential then to determine whether any *Dampness* is excessive and, if it is, the cause or response to other signs and symptoms. By so doing, practitioners can select required acupuncture techniques and help patients make necessary lifestyle adjustments. Recognizing *Dampness* through tongue, pulse and symptomatic presentations is more than a process of validating the various physical and mental health problems patients report. Taking time to explore the all-too-common triggers of *Dampness* and the lifestyle choices that may have precipitated the progression and establishment of *Dampness* helps practitioners and patients understand the cause of any associated microbiome dysbiosis and what changes may be successfully introduced to counter it.

Establishing the causes of *Dampness* and dysbiosis involves a thorough qualitative multi-sensory assessment of both mental and physical contributory factors. Chinese Medicine, especially acupuncture, is sometimes described as "an art," as practitioners in their interaction with each patient will make subtle adjustments and use, either consciously or unconsciously, some degree of intuition to guide them through multiple diagnostic insights (Kaptchuk 2000). Similarly, there is an "art of enquiry" that comes as practitioner and patient work together, helping decide which aspects of a patient's life may require further exploration, which may need to be revisited at a later stage, and what level of significance each aspect has. When a practitioner enquires about a person's life, this is not just fact-gathering. It allows a deep understanding of the patient's path to and experience of health and disease, including of the microbiome.

As noted earlier, the connection with *Earth* makes the circular emotional qualities and other correspondences of these *Organs* especially relevant to the assessment and treatment of *Dampness*. In Chinese Medicine, there are also other aspects that should be considered. *Dampness* is linked with "overeating," in terms of both quantity and quality. This could mean the overconsumption of rich foods, or an unbalanced diet heavily tilted towards processed and mass-produced foods devoid of nutritional content, laden with chemicals, additives, emulsifiers or pesticides. It could also mean too much refined sugar, artificial sweeteners and too few fresh, nutritious, natural and fiber-rich foods. The temperature at which food is served, time of day and frequency, including eating irregularly or too often, affect the digestive system. Use of medication, especially antibiotics, is known to disturb the natural rhythms of the digestive system through its powerful actions and these aspects should be noted. Any ill health, especially a history of parasites, viruses and bacteria which have *Dampening* effects both in the short and longer term with sometimes enduring symptoms of nausea and bloating, even extreme tiredness long after the apparent resolution of the infection, should be considered. Although *Dampness* may be key, it is still the case that other vital components of a healthy microbiome—*Qi, Blood, Jing* or *Elemental* balance—may be impaired and need consideration too. Simply put, using Oriental Medicine to unearth patterns of dysbiosis is a matter of understanding how the patient is, behaves and feels, and through this it will also become possible to work with the patient to define a microbiome-supportive lifestyle that draws on the knowledge of Oriental Medicine.

WIDER DIAGNOSTIC AND THERAPEUTIC CONSIDERATIONS

Fine-tuning the understanding of food and diet with the help of Oriental Medicine may therefore improve result outcomes. While specific precursors of dysbiosis may have been identified, there are some further areas for both practitioners and patients to consider.

Eating patterns

Since in Chinese Medicine the time of day and the season is linked to the health of the *Organs*, and can therefore be used to support them, eating a heavy meal late in the day and consistently eating out-of-season foods, such as chilled and raw salads during the cold winter months, can negatively impact on the energy balance in a person's system. The hours between 7am and 11am are known in *Five-Element* acupuncture theory as the *Earth* horary hours, when the key digestive Organs of the Stomach and Spleen are said to be "in charge." What this means in practice is that a person's digestive system is at its strongest at this time, making a well-balanced and nutritious breakfast especially beneficial. Breakfast should be thought of as a main meal, not a snack, and should provide a significant, if not the greatest, proportion of nutritional intake for the day.

Many will instantly recognize how this stands in sharp contrast to the everyday experience of living in a Western country where people frequently skip breakfast only to indulge in a heavy meal at the end of the day. When it is taken, breakfast is often a hastily eaten on-the-go meal, which involves low-fiber and high-sugar mass-produced foods and snacks, such as cereal bars and other processed breakfast foods. Taking time for a nutritive and wholesome breakfast is a daily ritual that can easily be implemented, as will be shown in the next chapter. Doing this allows a person to work with their internal body clock (see Appendix B), thereby aligning with the natural rhythms of the body, rather than going against them, to avoid unnecessary depletion.

The main emphasis should therefore be on achieving an appropriate energetic balance, in relation both to the time of day for meals, and also food selection and combinations. This notion is reflected in the breakfast consumed in many countries and regions across Asia, where the first meal of the day is designed to provide a nutritious combination of foods that is gentle on the digestive system and has either a neutral energetic effect or sweet flavor (supportive of the

Earth Organs), such as congee (rice porridge) topped with other ingredients including meat, fish, tofu or vegetables, or noodle soups served in a light broth or miso. Shifting the balance of nutritional intake to earlier on in the day will not only benefit the microbiome, it will reduce chances of overeating in the evening and going to sleep with a full stomach, thereby improving sleep, mood, memory, concentration, immunity and general vitality.

Movement

It is clear that individual lifestyles play a big part in determining the extent to which dietary factors affect or exacerbate any imbalance, and keeping active is as a result an essential part of supporting all aspects of health. In Oriental Medicine, sedentary lifestyles and lack of exercise typically contribute to the build-up of excess, whether *Qi stagnation* or *Dampness*. This lack of flow and movement can occur at a mental level too, with circulatory thought patterns, obsessive thinking or constant worry. Intense spells of mental focus can also contribute to patterns of stagnation. These features can quickly create an additional burden on the delicate balances that exist in the human body.

The need for regular exercise and to move frequently during the day to achieve health and balance in the microbiome and across body systems should not be underestimated. In Oriental Medicine, excessive sitting is said to impact negatively on the *Spleen*, which may weaken the digestive system. For desk-bound office workers, it may seem difficult to get around this, and the growing popularity of standing desks may only shift the problem to standing for too long, which is itself considered damaging to *Kidney* energy. Some degree of variation in posture and movement throughout the day can help rebalance energies and give space for body and mind to restore and repair. This could include something as simple as taking short breaks to stretch, walk and move. Such ideas are already embedded in many work settings. For instance, it is traditional in some workplaces and schools in Japan to start the working day with a series of stretching exercises known as *Radio Taiso* (García and Miralles 2016) that use all parts of the body. These are designed to improve mobility and flexibility, in both body and mind, and boost productivity too. There is also increasing awareness elsewhere around the world of the benefits of "deskercise" and workplace stretching programs to reduce the risk of injury and improve wellbeing too. Movement also encourages deep breathing, which is known to trigger both mental and physical changes, while helping move stagnation, making more generally such routines and breathing exercises beneficial to the microbiome.

Uniqueness

All individuals have their own norm, predispositions, likes and dislikes, much like the microbiome does. It is therefore important to recognize the extent to which a particular person is willing to embrace or is suited to specific foods that in theory are beneficial. Rather than dismissing the patient's views, practitioners must accept that there may be barriers—cultural, social, psychological or physical—preventing the patient from making the necessary lifestyle changes, and these should be explored with the patient to find suitable ways forward and be addressed where possible. In Chinese Medicine, there is a strong belief that the enjoyment of food itself is very important to be able to fully benefit from it. Food should create a sense of comfort and pleasure to maximize its positive health impacts. A more nuanced approach that helps foster a positive experience and emphasizes how certain foods should be prioritized provides a sustainable and achievable strategy. In other words, it is preferable for food and lifestyle principles to be laid out, so that changes can be implemented from the bottom up by patients themselves rather than dictated by a system.

While the considerations outlined so far—*Dampness, Qi, Blood, Jing, Elemental* imbalances, eating patterns, movement, breathing and individual circumstances—will give both patient and practitioner diagnostic and therapeutic tools to support health in the microbiome and beyond, there are two broad dietary principles and food groups that warrant special attention: fiber and fermented foods.

THE FIBER MANTRA

There is a clear consensus among experts in the field of microbiome nutrition that different types of dietary fiber are key to supporting a thriving gut microbiome and to ensuring associated health benefits. Although the word "fiber" is used as a generic term to refer to compounds that cannot be fully digested, and that are found in foods such as vegetables, fruits, legumes, nuts, seeds and grains, there are in fact many types of fiber. Each type of fiber has different levels of fermentability and functions, with all playing a crucial role in gut health, including providing roughage to speed up transit time, acting as prebiotics to encourage the growth and diversity of gut bacteria, or maintaining stability in blood sugar levels. Thus, eating a variety of these plant-based foods allows most types of fiber to be included in our diet to support a healthy microbiome from which many health benefits derive. It is worth noting that one type of fiber, found in raw oats and green bananas, known as resistant starch and which is fermented in the colon, is particularly important for the microbiome, as it feeds beneficial bacteria, releasing a by-product known as short-chain fatty acids (SCFAs). SCFAs help maintain the integrity of the lining in the intestines and prevent a "leaky gut," a condition that allows toxins and bacteria from the gut to enter the bloodstream, leading to inflammation. The connection between gut permeability and ill health has been dismissed by some doctors, while for others, including patients themselves, there is strong evidence of a causal link. It is nevertheless commonly agreed that a negative impact is at the very least theoretically possible and that the presence of a healthy gut lining is essential to preserve the proper functioning of the intestinal wall. SCFAs are generally believed to have significant positive impacts, having been linked to a reduction in the incidence of bowel disease, cancer, ulcerative colitis, Crohn's disease and antibiotic-associated diarrhea (den Besten *et al.* 2013).

Fiber-rich foods, which include fruits, vegetables and whole grains, are also nutritionally dense, providing a good array of essential vitamins and minerals. There is long-standing evidence that these foods form the basis of a well-balanced, nutritious and healthy diet, and reduce the likelihood of both acute and chronic diseases, including heart disease, diabetes and obesity. The connection between a high-fiber diet and improved health outcomes was first noted by Burkitt in the 1960s (Cummings and Engineer 2018) and has been documented in several epidemiological studies which have noted how nations and cultures with higher levels of natural dietary fiber intake have a lower incidence of these diseases. Interestingly too, a low-fiber diet has been associated with an increased risk of developing appendicitis, an acute inflammation of the appendix which usually requires the organ's removal. As the appendix holds a rich diversity of microbes that are used to repopulate the microbiome of the digestive tract when needed, its removal may not be as inconsequential as often believed. In any case, there are multiple negative knock-on effects resulting from an insufficient intake of fiber that go beyond increasing the risk of an appendicectomy. Inadequate levels of fiber may lead to a nutritionally deficient diet that starves bacteria found in the microbiome and disrupts the microbial ecosystem by reducing the number and diversity of microbes, while also decreasing the production of SCFAs. Keeping in mind a simple mantra of "fiber as often as possible" will do much to guide healthy food choices.

Fiber, it is worth emphasizing, plays a critical role in Oriental Medicine too, as vegetables and whole grains are considered the staples of a healthy and nourishing diet (Deadman 2016; Leggett 2014a, 2014b), ensuring overall balance and strength in both body and mind. Food therapy in Chinese Medicine, and in particular the concept of "food energetics," nevertheless provides additional principles of dietary balance, notably cooking methods and seasonal selections that can help personalize and therefore optimize the inclusion of fiber in a person's diet (see Appendix A). For example, raw foods have a cooling effect, steaming and boiling are seen as more neutral and easier to digest, stir-frying and stewing are more warming, whereas roasting has a stronger heating effect (Leggett 2014b). This qualitative variability can be used to help balance the energetic effect of a particular food—for instance, apples considered more cooling could be stewed (more warming) if a person is already showing signs of *Cold* (loose stools, lethargy). If there are indications of *Heat* in a patient (constipation, emotional volatility and restlessness), cooling strategies could focus on either the energetic effect of a particular food, or the method of preparation, or a combination of both, if a strong change is required. Understanding these influences may be especially useful to help patients gently reintroduce fiber-rich foods into their diets. Importantly too, these cooking methods reflect seasonal and climatic shifts and offers an important tool to learn to live harmoniously with the natural rhythms of life. During the summer months, more cooling food preparation may be favored over the autumn and winter months, when extra warmth is usually needed.

This alignment with the natural world and energies of the seasons can thus be used to guide food choices. By eating natural foods grown locally, diets effortlessly synchronize with the seasons and their changing climates. In many places, this will result in different types of fiber-rich foods being eaten more frequently at different times of the year. Across the temperate climates of the northern hemisphere, the availability of locally grown fresh fruits and vegetables often follows the color patterns of the *Five Elements*: a more abundant supply of greens is normally available in the spring, bright red fruits such as strawberries, raspberries and cherries in the summer, yellow or orange-colored vegetables, especially squashes (butternut, pumpkin) and root vegetables grown deep in the earth in the late summer, and white vegetables such as cauliflower, mushrooms, turnips, onions and kohlrabi in the autumn. The winter months are when many crops are put to "rest" and this is a time when other types of fiber that have been stored and preserved, such as grains and cereals (barley, millet, oats), dried fruits and vegetables from the sea such as kelp, come to be used, even if many hardy vegetables, such as Brussels sprouts, cabbages or kale, can survive the harsh winter conditions and be harvested at this time of year. Each season is associated with a distinct flavor—sour (spring), bitter (summer), sweet (late summer), pungent (autumn) and salty (winter)—which reflects the seasonal movement and action of distinctive energetic shifts. This directionality can be used to influence desired energetic outcomes in relation to a person's *Yin–Yang* balance and enhance the therapeutic effectiveness of nutrition (Leggett 2014a, 2014b; Pitchford 2002).

This *Elemental* dietary approach not only helps understand better how beneficial foods, such as those rich in fiber, can be incorporated in a meaningful and harmonious manner to nourish the balance in the microbiome, it provides an anchor around which menus and dishes can be designed (see Chapter 5). Before looking into these applications, it is important to consider one other group of foods that plays a very special role in finding and maintaining a healthy balance in the microbiome, and which has become somewhat of a sensation in health circles: fermented foods.

FERMENTED FOODS

Eating out-of-season foods is hard to avoid in today's world, as food stores have a year-round supply of almost all fruits and vegetables, including many non-native varieties. There was a time, however, when preservation methods were needed to minimize food wastage and shortages. Drying grains helped deal with the food scarcity of the colder months, but so did other preservation methods, notably fermentation, which provided, before artificial refrigeration existed, a vital way to both reduce spoilage, especially during the warmer months, and avoid the risk of ingesting harmful pathogens. Fermentation was used not only to preserve foods but to change and improve flavors, textures and even the effects of foods when consumed (notably alcohol), simultaneously allowing humans to unwittingly deepen their symbiotic relationships with bacteria. Fermentation is certainly one of the main actors in the entangled story of the co-evolution of humans and microbes, and indeed their enduring friendship. This is because fermentation is a naturally occurring phenomenon that creates transformative energies that are intrinsic to life.

There is a long history therefore of using fermentation in the preparation of foods and drinks, to preserve and improve taste and usability. Today fermented foods and drinks such as kimchi, sauerkraut, kefir, yoghurts and kombucha are back in fashion as the health merits of their various strains of live cultures are increasingly well understood. The natural fermentation processes used in these foods gives them a distinctive and authentic flavor, making the experience of eating a sourdough made from "wild" yeast very different to eating a plain loaf of mass-produced bread. There are also many everyday foods that have been fermented as part of their preparation, which most people are unaware of, including different types of bread products, cheeses, vinegar, wine, beer, miso, soy sauce, traditional salamis and other dried sausages. Even cocoa beans need to be fermented to give chocolate its aroma. Around the world fermented foods are often deeply embedded in local cultures, customs and the availability of native crops, fish or meat. There is a very special bond between humans and fermented foods and a cultural authenticity that gives them a fundamental place in today's world of nutritionally poor diets that focus too much on convenience and calorie counting.

Fermented foods are considered indispensable in the context of the microbiome because they contain probiotic micro-organisms as well as digestive enzymes. It is essential to note that micro-organisms from fermented foods can only be maintained for a relatively short period of time within the gastro-intestinal tract, due to the highly acidic environment of the stomach. However, both in-vitro studies and randomized controlled studies suggest that the physiological changes that take place as a result of the ingestion of these microbes are significant since they activate, regulate and modify various functions including cardiovascular, immunological, metabolic and digestive functions (Dimidi *et al.* 2019). Given the powerful, albeit transient effect of fermented foods, the best way to utilize them is on a regular basis, in a balanced way and within a varied diet.

From a Chinese Medicine point of view, the energetic effect of fermented food will depend on the ingredients and the precise process of fermentation. That said, fermentation often creates a distinctive "sour" flavor, which may be connected to the *Wood Element* and its associated *Organs, Liver* and *Gallbladder*. Interestingly, Katz (2012) explains in his enlightening history of fermentation how a fermented condiment widely used in ancient China known as "Jiang" was compared to a military general who was said to draw out the poison from food. Not only does this analogy depict how spoilage-causing bacteria are unable to take hold in the acidic environment produced by fermentation, *Liver* in *Five-Element* acupuncture is similarly known as the "General of the Army," the Official in charge of strategy and planning. The beneficial micro-organisms found in fermented foods, with their distinctive "sour" flavor, may be considered supportive of the work of the *Liver*, and significantly of the smooth flow of *Qi*. This parallel also helps explain why fermented foods, which have already been through a process that breaks down sugars and starches, are more easily digested. This perspective on fermented foods provides several therapeutic applications to support patients, and may include other *Elemental* principles, for instance ensuring the inclusion of fermented foods during the springtime in particular, or to support detoxification. There are of course other energetic actions that may be associated with different fermented foods, depending on flavors and ingredients. For example, a salty and sea-like flavor is often noticeable in foods such as kimchi and may thus be seen as supportive of the energetic balance of *Kidney* and *Jing*.

Such *Elemental* associations are summarized in a short guide in Appendix A, which also outlines how preparation, cooking methods and seasonal availability influence energetic effects and can inform the overall nutritional balance of a diet. It is important to reiterate that the guide is not intended to be prescriptive and can be used for reference, to plan energetically appropriate meals for each individual or season, and more generally to explore dietary options in a way that helps bring together traditions of Oriental Medicine and gut health. Crucially too, foods that are available year round do not necessarily need to be restricted to a specific season, even if they have an *Elemental* association. Moreover, those connected to the *Earth*, and therefore supportive of the *Spleen*, provide a good basis throughout the year and are generally well tolerated. This explains why foods like congee (rice porridge) are often eaten across many parts of East Asia when convalescing, or when a person's constitution is especially weak.

CONCLUSION: FINDING THE RIGHT BALANCE

Fiber and probiotics, such as those found in fermented foods, play a chief role in the health of the microbiome, and their dietary use can be further tailored to the needs of each individual. By considering *Elemental* affinities, practitioners and patients can easily identify times when these may be prioritized and may be especially beneficial. The selected foods found in Appendix A, and which are organized around the *Elements*, include food types that positively contribute to the balance of the microbiome. It is noteworthy that some of these foods include not only fermented and high-fiber foods but also foods and drinks such as olive oil, tea or coffee, and other food groups, including omega-3-rich foods such as salmon, which are favorable to gut health. Many plant products, for example red wine, berries, olives and herbs and spices such as mint and cloves, also have high levels of polyphenols. Polyphenols are micronutrients that are broken down by bacteria in the colon and are thought to offer protective effects against some degenerative diseases (Manach *et al.* 2004).

While the considerations discussed in this chapter provide simple core principles that can help guide patients and practitioners towards a microbiome-supportive diet, it is essential to remember that finding the right balance for each person relies on a mixed methodology in which food choices are integrated into one's life in a manner that is enjoyable, truly nourishing and congruent with oneself. To understand better how this may be achieved in practice, let us consider next these principles in context.

Chapter 5

Food for Thought and Practice

No matter how sound our theoretical understanding of the microbiome and Oriental Medicine might be, it is only through the application of this knowledge that as both practitioners and patients we can truly experience the benefits of the positive changes that we consciously and unconsciously make to the way we live and importantly to the way we eat. As a first step towards fully recognizing how this combined knowledge of the microbiome and Oriental Medicine can be integrated into everyday life and eating practices, this chapter explores recipes that illustrate the different ways in which *Elemental* qualities are imparted through food to nourish and harmonize *Organs* while supporting the microbiome too.

ELEMENTAL NUTRITION

From the descending movement of late autumn and the winter months to the outward push of spring towards the peak of summer and beyond, the seasonal cycles and transitions bring energetic shifts which can readily be seen in the changing qualities of the living world. These changes mark different points of transition in the interrelationship between *Yin* and *Yang* on an energetic spectrum which stretches between winter—the most *Yin* phase of the seasonal transformation, characterized by a slow, contracted, quiet, dark and cold energy—and summer—the most *Yang* stage, emblematized by a fast, light, bright and hot energy. These adaptations dramatically alter landscapes as we move across the seasons, and significantly too define many aspects of human health. Oriental Medicine sees the human body as a reflection of the natural world and as such, good health is dependent on living in harmony with the laws of nature. This is why appropriately designed seasonal nutrition and diets are used therapeutically in Chinese Medicine both to address specific health concerns and as preventative medicine.

This alignment with nature is especially relevant when it comes to supporting the health of the microbiome, since the story of the microbiome can be traced back to the very beginnings of life. It is a tale which weaves together the evolutionary paths of humans and microbes within the natural world and reminds us of the need to maintain a positive connection between ourselves and our environment. The principles of harmonious eating which underpin both Oriental Medicine and gut-health diets provide a simple and easy-to-follow approach to nourishing the balance of the microbiome and supporting more broadly physical health and mental wellbeing.

Understanding the seasonal patterns of the *Five Elements* and the daily energetic changes that occur every two hours, according to Chinese Medicine's Body Clock (see Appendix B), allows humans' internal workings to harmonize with their environment and maximize a person's ability to harness the energetic potential afforded each day. From a dietary point of view, this means making the most of local and seasonal foods, taking into consideration the climate within which a food has been grown and is being eaten, the temperature at which it is being served, and the way in which it has been cooked, so that eating takes place in a manner that is consistent with the energy of the time of day and year. In the wintertime, more warmth is needed, while in the summer, cooling foods are most appropriate. As highlighted in the previous chapter, other seasonal associations—whether in terms of flavors that will be most beneficial, colors or specific food types—offer helpful pointers to create internal and external harmonization. With the *Earth Organs* being at their strongest in the morning, eating a nourishing breakfast that strengthens all the *Elements*, followed by a balanced seasonal lunch and a lighter dinner will help respect the strength of all *Organs* across the day and seasons.

Meals should be a relaxing and enjoyable moment, and making time to prepare and eat a dish, creating a nice space in which each meal is taken, engaging fully with the flavors and textures of the foods, will make it easier to achieve this and to follow other recommendations given by Oriental Medicine practitioners, such as chewing well and not rushing meals. What may seem to be a subtle change or inconsequential aspect of our eating habits is, energetically speaking, considered influential in Chinese Medicine. For instance, a delicately presented dish is seen as nourishing to the *Lung Organ*, variety in our diet to the *Kidney*, freshness to the *Liver*, while sharing a meal with loved ones can bring additional nourishment to the *Heart* (Leggett 2014a). The *Spleen* for its part is positively affected by having a deep sense of comfort and feeling cared for (ibid.), perhaps taking us back to cherished moments in our childhood.

GETTING STARTED

Knowing how or where to start to improve the health of the microbiome will depend on individual circumstances. If a patient is currently seeing a practitioner of Oriental Medicine, the practitioner will have a very detailed understanding of any imbalances, deficiencies, inherent *Elemental* strengths and weaknesses, and other signs and symptoms, which may be suggestive of very specific dietary guidance. The Traditional Diagnosis undertaken in *Five-Element* acupuncture at the patient's first appointment usually points to certain issues that may guide such recommendations. It is key to ensure these are tailored to the patient, especially in cases of extreme depletion where it will be essential to consider what demands any dietary changes may be making on the patient.

Both practitioners and patients would probably agree that what is needed to ensure the sustainability of dietary changes is an approach to eating that is easy to grasp, effortless to follow and that fits into a person's life—whether from a practical point of view or in relation to personal likes and dislikes. In other words, each individual needs to find the foods and meals that work for them, so that "eating well" can become a way of life rather than a strict diet. What follows is intended therefore as a guiding tool, which will need to be adapted and built upon to reflect an individual's own personal circumstances, either with the help of a practitioner or through an objective self-assessment.

In addition to the *Elemental* dietary considerations already discussed, some basic principles can be kept in mind to facilitate the adoption of this microbiome-balancing approach to eating. On a daily basis, ideally around three quarters of the food intake should be of vegetables, fruits and grains, with the final quarter made up of eggs, dairy, nuts, fish and meat products. If the meals are satisfying and healthy, there will less likely be a need for snacks, which tend to disrupt digestion, and negatively impact on the microbiome's profile, especially if they are sugary. The bulk of fluid intake in the form of water served at room temperature and herbal teas should be taken between meals, while small amounts of water or other drinks can be taken at, or around, mealtimes. This could include in moderation tea and coffee, and if desired, from time to time and with an evening meal, a small glass of red wine, which has been associated with many health benefits, including for gut microbial diversity (Le Roy *et al.* 2020).

HOW EVERYDAY RECIPES NOURISH THE BALANCE

To highlight the varied ways in which Oriental Medicine can teach us how the microbial balance can be supported at a physical and mental level, let us explore five everyday recipes. The recipes have been selected as examples of dishes that demonstrate the principles discussed throughout and show us how foods have a story, with particular *Elemental* qualities that nourish us and our microbiomes.

Building the foundations: Kefir-Bircher

A kefir-based Bircher, or "overnight oats" as it is often called, is a good example of a microbiome-nourishing dish that can be seen in Oriental Medicine to regulate the health of all *Organs*. Based on the famous Swiss *Birchermuesli*, a kefir-based Bircher provides a simple yet nutritious breakfast that needs just a few ingredients—grains, such as oats; fresh, dried or cooked fruits; nuts; seeds; and fermented milk (or live yoghurt), either dairy or plant-based. These ingredients are all combined in a bowl and left to rest for 8–12 hours. It is a dish that can be easily adapted, as ingredients can be chosen to reflect individual needs, preferences and seasonal availability. The original Swiss recipe was popularized by a Swiss doctor, Dr. Bircher-Benner, at the start of the twentieth century and was inspired by the inhabitants of the mountainous regions of Switzerland who ate a humble yet wholesome diet, rich in natural and highly nutritive foods farmed through traditional methods that were to all intents organic and biodynamic (Patrimoine Culinaire Suisse 2020). The mountaineers' diet was also extremely efficient, allowing for very little food waste. For example, the organic apple used in the mountaineers' precursor to Dr Bircher-Benner's recipe was grated in its entirety to include skin and core. These details are of considerable significance in the context of microbiome nutrition, as recent studies have found that the core of an apple is a rich source of microorganisms, and that organic apples yield a greater diversity of beneficial bacteria and fungi than regular apples (Wassermann, Müller and Berg 2019), making the Swiss mountaineers true front-runners of gut health. Dr. Bircher-Benner's philosophy aligns very closely to both Oriental Medicine and gut health, as he strongly believed in the healing power of food. He further recognized the specific energetic and health potential of food which results from the conditions in which food is made, including the sun, light and air that goes into growing crops.

The combination of raw oats soaked in live kefir or yoghurt, as well as the fruits, nuts and seeds, make Bircher a deeply satisfying breakfast that is quick to prepare, easily digested and will keep energy levels well stocked up until lunchtime. It is a real treat for the microbiome too. Raw oats are high in resistant starch, while the fruits provide a rich source of different vitamins and prebiotics. Kefir has long been hailed for its wide-ranging health benefits and includes a range of strains of bacteria and yeasts, including diverse species of *Lactobacillus* (Slattery, Cotter and O'Toole 2019). The specific cultures of the kefir will of course vary depending on the manufacturing process. Homemade kefir usually provides a wider variety of bacterial strains. That said, ready-made kefir contains many beneficial probiotics and can be used for convenience and to further facilitate the preparation.

The basic ingredients used in a Bircher, when looked at through the lens of Chinese Medicine's food energetics, are nourishing to all *Yin Organs*—*Kidney*, *Heart*, *Spleen*, *Lung* and *Liver*—and together have a powerful balancing effect. Oats are supportive of all *Organs*, even if their light, sweet flavor gives them a direct connection with the *Earth Element* and, when eaten raw, with the *Fire Element*. They provide a gently warming and grounding energetic effect too. Other grains such as millet flakes, supportive of *Kidney*, or wheatgerm, which benefits *Heart*—can be used instead to suit individual require-ments. By including a mix of dry and fresh fruits, different energetic effects can be harmonized. Dry fruits have a warming quality in Oriental Medicine and will be especially useful in the winter, and additionally provide the means to offset the effect of more cooling ingredients, such as seasonal raw fruits. Warm stewed fruits can also be added at the time of serving, which can help compensate for any excessive cooling energy from having kept the Bircher to rest in the fridge.

The choice of nuts and seeds that can be added is endless, and can be either whole, chopped or even in a paste such as peanut butter, almond butter, tahini or cashew butter. It is essential to remember that nuts can have a *Dampening* effect if used in large quantities, and moderation is therefore key. Patients and practitioners should still consider individual susceptibilities to *Dampness*, lifestyle and energetic patterns to guide the inclusion of nuts. Some people will tolerate lashings of peanut butter much better than others. Spices provide a helpful additional tool to help adjust energetically the dish and can include any spice suitable for baking, such as cinnamon (*warming*), cardamom (which reduces *Dampness*) or even mixed spice (*warming*). Bircher makes an uncomplicated yet deliciously satisfying breakfast or lunch. In Switzerland, it is still eaten as a meal at any time, and from a Chinese Medicine point of view can be suitable all year round with appropriate modifications to ingredients and preparation methods.

The path to growth: Sourdough

Sourdough is more than food, and much more than bread too. It illustrates many concepts central to understanding the dynamics of the microbiome's equilibrium and the effect of specific foods across different systems of medicine. Bread is often considered problematic in Oriental Medicine, as it is generally seen as *Damp*-forming. As pointed out in Chapter 4, *Dampness* is consistent with patterns of gut dysbiosis and is frequently associated with weight gain, digestive problems and fatigue. While some breads may contribute to such problems for some patients, bread can, according to both Eastern and Western nutrition science, also be included in a balanced and healthy diet. Much depends on how often it is eaten and importantly how the bread has been made and its ingredients, as artisanal or homemade sourdough which uses organic whole grain and wild fermented yeast brings many health benefits.

Whole grains are vital nutrients in both Eastern and Western systems of medicine, due to their *Qi*- and *Blood*-nourishing qualities (Leggett 2014a, 2014b), high vitamin and mineral contents and the role they play in protecting against fatty liver disease, heart disease, stroke, obesity and diabetes (Harvard T.H. Chan School of Public Health 2021b). Moreover, the high fiber content of whole grain makes it a supportive ally of the microbiome and its metabolic and anti-inflammatory effects. There are also many arguments in support of opting for organic whole grains. While improved taste may be a subjective assessment, it is undeniable that organic whole grain is pesticide free and therefore both environmentally friendly and supportive of farming practices that align to *Taoist* principles of harmonious living with nature.

The wild fermentation of the starter known as "The Mother" allows for the growth of a range of micro-organisms that are used in the preparation of the bread, including yeast, which enables natural rising, and lactic acid bacteria, which give the sourdough its distinctive flavor and maintain its shelf life. While these live bacteria do not withstand the high cooking temperatures, they do change the structure of the flour during the preparation and baking as the gluten content is reduced. For this reason, sourdough is much easier to digest than most fast mass-produced breads. Fermenting the bread in this way has additionally been found to enhance fiber content and phenolic compounds (Taneyo Saa *et al.* 2017), which are supportive of the health of the microbiome. A home-baked sourdough bread nourishes the microbial balance by other means too. The preparation of the bread involves kneading the dough with bare hands, giving the skin microbiome exposure to a variety of micro-organisms, and vice versa. While this exposure is only temporary, much like the transient presence of bacteria in the gut from eating probiotic-rich foods (see Chapter 4), it is conceivable that this interaction may trigger further changes that have a longer-lasting effect. Furthermore, due to this "bidirectional exchange of micro-organisms" (Reese *et al.* 2020), there is a somewhat unique microbial profile to each sourdough loaf which is linked to variability in the ingredients and environment, including the baker's skin microbiome. This uniqueness thus provides an additional avenue to support a greater diversity of microbes in the microbiome.

Another way in which preparing and eating sourdough has a harmonizing effect is through the emotional qualities it may generate. While this idea is much more closely aligned to Oriental Medicine's food energetics, its relevance to microbiome science can also be seen through the gut–brain axis. The emotional affordance of sourdough comes in part from the preparation, which extends over a period of several days. Partaking in such activities, where it is important to allow "time" for changes to happen between each successive step in the recipe, provides a powerful lesson in learning to be patient. Preparing a sourdough is energetically akin to the transformation that occurs from the point of germination until harvest time and allows us to experience the natural rhythms of growth as we smoothly work through the systematic steps that bring the project to fruition. Making a sourdough requires appropriate and timely attention to be paid each day to its preparation. There is something of a military operation involved in doing so: respecting the timing of each step, keeping an eye on progress and engaging at precise points in kneading to punch down the dough while following a clearly defined strategy. The exactitude of the operation is noted by Katz in his famous book *The Art of Fermentation* (2012), in which he describes how precise measurements are needed too. Katz also describes the importance of patience and calm while we wait for the reward: even when the sourdough is seemingly "ready" after being cooked, it remains essential to let it cool down a little, as the very final bit of cooking takes place when the loaf is out of the oven, giving it its unique flavor and texture.

Sourdough preparation is clearly reminiscent of the healthy

workings of *Liver* in *Five-Element* acupuncture, the controller of time which ensures the smooth flow of *Qi*. Its sour flavor and upward movement as the bread rises mirrors the energetic direction of *Liver's* associated *Element Wood*, while the *Ke* cycle discussed in Chapter 3 goes some way to explaining the digestibility of sourdough compared with other breads, and its ability to create a stabilizing sense of homeliness. The smell of freshly baked bread has a very grounding and comforting effect, evoking a coziness which is often said to be triggered by memories of childhood. These aspects align well with the energetic of a healthy *Earth Element*, and it is fitting that the starter culture of a sourdough is called a "Mother," a figure in Chinese Medicine associated with the *Earth Element*. Interestingly, starter cultures are often gifted among family members or friends, a gesture that itself echoes this dynamic between the benevolence of the *Wood Element* and the nurturing effect of *Earth*. It is a dynamic energetic exchange that supports overall health and wellbeing and shapes the health of the microbiome too. This healthy relationship between *Wood* and *Earth* is perhaps best emblematized in the acupuncture point GB-24 "Sun and Moon," which represents the balance between the fundamental energies of *Yang* and *Yin*. A crossing point of the *Gallbladder* and *Spleen* channels, GB-24 is often used to address "*Damp-Heat*," which, as noted in Chapter 4, is closely associated with dysbiosis in the microbiome. The point's indication for *Rebellious Qi* further supports its use for symptoms of gut dysbiosis as it aids appropriate flow of *Stomach Qi* and resolves problems such as acid reflux, nausea and bloating. So, it may be seen that sourdough affords similar qualities of growth and balance.

Feeding the fire: Summer tabouleh

The abundance of locally grown fresh fruits and vegetables tends to make our diets healthier in the summer, and this is also a time of year when the warmer temperatures allow for the appropriate use of more cooling foods. Maintaining an energetically balanced approach to nutrition across the seasons is key in Oriental Medicine and benefi-cial to the microbiome's health too. It can nevertheless be difficult for those whose digestive systems have been weakened, and for whom the consumption of foods that have cooling and moistening effects is problematic, to include such seasonal foods, and raw vegetables in particular. In the summer months, it is important therefore to try find ways of keeping "cool" while not exacerbating *Cold* or *Damp* conditions. This is where fermented vegetables can be particularly helpful, allowing Eastern and Western approaches to harmonious eating to be reconciled. These can also easily be combined with grains, such as the bulgur wheat used in tabouleh, to make a light and well-balanced dish.

The tabouleh salad, which finds its origins in the Middle East, provides a good example of a dish that can be used to combine all appropriate *Elemental* qualities for the summer months, while ensuring the microbiome's balance. A tabouleh made from bulgur wheat, seasoned with fresh herbs, lemon, black pepper and olive oil, topped with fermented vegetables, and for example roasted and cooled peppers, illustrates the use of ingredients that support the *Fire Organs—Heart, Small Intestine, Pericardium* and *San Jiao—* that dominate during the summer months and are central to many aspects of our everyday health. Imbalances in these *Organs* can take many forms, often showing up with distinctive symptoms including profuse or spontaneous sweating, pallor, heart-related problems, breathlessness, inflammation, anxiety, low mood and depression, intense dreaming, insomnia, social phobias or withdrawal, confusion and mania, to name a few. The *Heart* or *Sovereign Ruler* has of course a particularly significant role in overall health and emotional stability in Chinese Medicine and keeping it well, mentally and physically, is critical to avoid unpleasant symptoms. The *Heart's* connection to the tongue can give rise to very specific patterns of disharmony, such as incessant talking or difficulties communicating.

It has already been pointed out that the *Heart's* connection in acupuncture theory with the *Mind* and *Small Intestine* makes its influence on the health of the gut microbiome and vice versa significant. Interestingly, the tongue is itself a hotbed of microbes that are ingested through saliva and link the microbiome of the oral cavity and gut, and is key to the overall balance of the mouth micro-organisms that influence cardiovascular health (Nelson-Dooley 2019). Feeding the *Fire Element* is therefore an important part of the story of how to nourish our inner balance, not only because of this interrelationship with the microbiome, but also because it is energetically linked with *San Jiao*, which is directly involved in maintaining homeostatic mechanisms, including those pertaining to digestion. Creating harmony in the *Fire Organs* is both nourishing to the soul and to thriving microbial communities, and a tabouleh with fermented vegetables provides an everyday dietary input to supporting this goal.

Nourishing the Earth: Seasonal soups

The *Earth Element* in Chinese Medicine is an anchoring, sooth-ing and caring energy, which provides a homely, comforting and nourishing feeling. Eating in a manner which allows a sense of deep contentment is both the very essence of the *Earth Element* and further supports its functions and *Organs* (Leggett 2014a). The virtuous circle of eating to support our *Earth Organs—Spleen* and *Stomach*—and being supported by them and the energy they offer, are essential for our health and wellbeing and a thriving microbiome. Soups have a natural affinity to the *Earth Organs* and to the micro-biome, being made with a wide variety of vegetables that provide warmth and fiber-rich meals that can be given different flavors, textures and energetic effects. Soups are easy to prepare and can readily be adapted to match seasonal food availability, personal preferences and Chinese Medicine's various therapeutic emphases. There are many variations and combinations possible, and by using the classifications outlined in Appendix A, seasonality and energetic effects can be taken into account.

In addition to the fiber from the vegetables, soups can include many ingredients that contribute to the health of the microbiome, including fermented bases such as miso (made from fermented soybeans) and bone broths. Broths have been traditionally used across many cultures for their healing properties and are considered by some health practitioners today as beneficial to the mucosal lining of the intestinal wall and to bacterial diversity. Soups are usefully included as part of a varied diet and are especially nourishing for those seeking to preserve their energy, whether because they are diagnosed with a *Cold* condition, live in a cold climate or want to support microbial variety in an easily digestible way. Although they are similar to salads in terms of their versatility and broad use of vegetables, it is worth pointing out that soups provide, from a Chinese Medicine perspective, a highly balancing meal that is especially useful for those with a weak digestive system, since they circumvent the *Cooling* effect of the raw vegetables often used in salads.

The lightness of breathing: Baked tofu and vegetable rice bowl

Life depends on breathing in both Eastern and Western systems of medicine and on the work of the lungs. As a baby enters the world, the first breath is taken within a few seconds of birth, clearing the fluid-filled lungs and triggering momentous changes that enable life outside the womb. The lungs' responsibility for the exchange of gases that allows oxygen to be brought in and carbon dioxide waste to be removed further underlines the lungs' critical role and their strategic location as a first line of defense against pathogenic invasions. This means that the lungs inevitably play an essential part in microbial health too, as the gut–lung axis suggests. The lungs sit at the crossroads of many functions in Western medicine, and maintaining a healthy *Lung* energy is similarly vital for many reasons in Oriental Medicine.

According to Chinese Medicine, *Lung* is associated with *Qi*, as well as respiration and skin, and its depletion may lead to problems such as impaired or extreme sense of smell, nose bleeds, rhinitis, as well as low or weak immunity, inappropriate sweating, cold limbs and poor circulation. Disruption in its paired *Organ, Large Intestine*, may manifest physically as constipation or diarrhea, or as mental symptoms resulting from the build-up of "rubbish," including lack of self-respect, an inability to focus, obsessiveness with cleanliness or a certain fanaticism about order and precision. Health in the *Metal Element* leaves us feeling clean, pure and fresh in body, mind and spirit. Supporting this *Element* through food therapeutics is important therefore to support the vitality of *Qi*, a well-functioning immune system and health in associated *Organs* and key microbial sites, including the skin and colon. A good way to do this can be by eating meals that are pure, enticing, beautifully flavored and perfectly balanced. A baked tofu and vegetable rice bowl can do just this. Ingredients for such a dish can be especially apt at nourishing and balancing the *Metal* energy. Tofu, mushrooms, ginger, rice, onion, radish and cooked lettuce all have an affinity with the *Metal Element* and can help tonify *Qi* and resolve *Dampness*. The flavor is pungent, benefiting the respiratory system. The cooking methods used, which include boiled rice (*neutral*), stewed lettuce and radishes (*warming*) and baked tofu (more *warming*), allow the patient's energetic correspondences to strengthen in a gentle and sustainable manner. It is also a dish that echoes the simplicity, lightness and delicate nature of the *Metal Element*.

CONCLUSION: THE WAY TO SYMBIOSIS

These recipes have sought to show the ways in which both *Elemental* energies and the microbiome can be supported, while maintaining a harmonious and symbiotic relationship with our environment and the natural rhythms of the seasons. The examples of the dishes described here demonstrate that there are many nutritional qualities to foods that go beyond their material aspect and that the exact definition of "eating well" is highly contextual and relative to every person. Far from being a strict dietary protocol, nourishing the microbiome's balance is, fittingly for Oriental Medicine, "a way" of being, effortless and naturally attuned to the unfolding of life. Thus, it is a dietary approach which forms part of the *Microbiome Way* to which we now turn.

Chapter 6

The Path to Harmony

THE MICROBIOME WAY

O ur exploration of the microbiome through the lens of Oriental Medicine has shown that there is much for both patients and practitioners to gain from integrating the wisdom, knowhow and practice of Eastern and Western therapeutic approaches. Doing so gives rise to new opportunities and a fresh perspective on healing, preventing disease and maintaining good health. So far, our focus has been predominantly on how food and nutrition underline a therapeutic cross-over between Eastern and Western medicine which may be used as a platform to inform and personalize dietary choices. Oriental Medicine has, however, much more to offer than the, albeit highly valuable, nutritional calibration tools of food energetics. As a medico-philosophical system, it contains many important lessons that can be used to develop a natural and easy microbiome-friendly way of life.

Drawing on Oriental Medicine's core values, this chapter outlines five simple everyday practices to nurture the art of harmonious living. More specifically, it is suggested that paying attention to these five key principles not only enhances overall health and wellbeing, it can also positively contribute to the health of the microbiome too. Once again, this approach is not intended to be dogmatic and seeks instead to present a practical and explorative instrument to tap into our innate healing powers so that each and every one of us can find, with our microbiome, our path to health and longevity. This is the *Microbiome Way*.

THE MICROBIOME WAY

The "way" is used here to emphasize a principle at the heart of Chinese Medicine which encapsulates the profound, timeless and universal teachings derived from Lao Tzu's *Tao Te Ching* (see Mitchell 1999), in which there is a natural and somewhat effortless order in which life unfolds. The *Tao Te Ching* (The Book of the Way) is a guide to the art of living and, if applied in today's world, affords an openness to embracing all possibilities of life simply by renouncing preconceived judgment, relentless desires and the obsessiveness that modern life so often fuels. While some patients may struggle to see how such features could influence their everyday health, practitioners of Oriental Medicine are only too aware of the troubles these emotions can cause, whether in the form of an *Elemental* imbalance, *Blood* and *Qi Stagnation*, *Blood Stasis* or unnecessary depletion of their *Jing*. These energetic impairments can weaken healing or manifest in a variety of physical or mental disorders, including pain, infertility, fatigue, stress, anxiety, depression, insomnia, tinnitus, digestive problems, auto-immune conditions and so on.

The *Microbiome Way* gives practitioners a tool to help patients understand factors impacting on their health in a manner that aligns with the principles of Oriental Medicine, while also allowing for a fuller and richer experience of the transformative potential of the acupuncture treatments themselves. The practices are organized around the energetic qualities and possibilities afforded by each of the *Five Elements*. By so doing, practitioners and patients can thread together the everyday *Elemental* steps that can be taken to establish a congruence between personal and environmental forces to nourish balance and lead fulfilling lives. The five practices below are designed as an instrument to support patients and practitioners, and to bring a new experiential dimension to Oriental Medicine and microbiome science.

Practice no. 1: Cultivating the self

Working in harmony with the microbiome should be seen as an innate process, where ecological stability is born from the laws of nature. This understanding is important as it moves away from the idea of "mastering" the microbiome, recognizing instead how it is an integral part of human life and the *self*. It provides also an essential first step in improving the health of the microbiome by emphasizing the principle of self-awareness. Some patients, even when they are looking for help to improve their health or a particular condition, may see their lifestyles as an acceptable normality and may even be resistant to the idea that there is a connection between how they lead their life and their health. Whether it is daily stresses, emotional volatility, restrictive diets, excessive physical activities, unnatural aesthetic treatments or too much alcohol, the normative discourse surrounding many everyday habits and lifestyles has blinded us to the true impact of these practices and to the extent to which they are distancing us from who we really are and the purpose of our lives.

Self-awareness arises from a strong mind–body connection and allows proper functioning across body systems. It influences the health of the microbiome through communication pathways, such as the gut–brain axis. There are many ways to support a positive link between mental and physical energies, both through therapies such as acupuncture and through exercise, whether walking, running, cycling, swimming, *Yoga*, *Tai Chi* or *Qigong*. The increased flow of *Qi* throughout body and mind that comes from this improved connectivity is significant and may be used to help address stagnation and associated patterns of dysbiosis.

Such self-awareness facilitates deeper self-reflection too, from which grows an appreciation and acceptance of who we are and our place in the world. In Chinese Medicine this balanced state of being emanates from a healthy alignment of the *Kidney* and *Heart*, an axis which represents the balancing of *Water* and *Fire*, and the fundamental forces of *Yin* and *Yang*. Diagnostically and therapeutically, acupuncture can usefully identify and support patients with this alignment, as there are many different acupuncture point combinations and approaches to rebalancing the forces of *Yin* and *Yang* and to re-establish more specifically good communication between *Kidney* and *Heart*, whether by treating the lower and upper parts of the body, posteriorly and anteriorly, through selected *Yin* and *Yang* channels, or specific points, notably *Kidney Chest* points and *Bladder Back Shu*.

Connecting with the *self* provides a potent tool also to reinstate a person's trust in his or her inborn ability to heal and nimbly progress through life. Documented tales of healing (e.g. Rediger 2020), even from severe illnesses, provide powerful testimonies of the influence of the mind–body connection on health outcomes and how even something as subtle as a shift in perspective may change the course of a disease or health condition. I once discussed informally with a consultant surgeon at a local hospital this very question and how significant patients' mindset could be for their health. Interestingly, he remarked that he had noticed, from anecdotal evidence, how cancer patients who believed they would get better seemed to have more positive outcomes compared with those patients who had a more pessimistic outlook. Such observations suggest parameters of healing that go beyond the physical realm, and which are of course highly relevant to understanding how to support microbial health. Non-material variables of health are deeply embedded in the principles of Oriental Medicine as the emotion, virtue and *Spirit* associated with the *Organs* bring with them an energetic quality and potential to empower all aspects of existence from which the "*true self*" can arise. A strong and peaceful connection with the *self* and its self-referential framework with the outside world can mobilize appropriate responses from within that may create movement and transformation necessary for health and healing.

Some patients may find it challenging to connect with their "*true self*," since they may have been disengaging from it for a long time, either consciously or unconsciously. In such situations, appropriate use of specific acupuncture points and protocols such as *Seven Dragons*, used in *Five-Element* acupuncture, will be especially beneficial. Such treatments can aid self-understanding and the awareness that allows patients to acknowledge and let go of past regrets, mistakes, shame or blame, so that they can recognize those moments, no matter how brief, where peace and happiness prevailed and in which the free expression of their *self* was possible. Making this connection with the *self* provides a grounding and invigorating experience for patients as it brings them to the present moment, awakens their senses, nurtures their growth and provides stability at body, mind and spirit levels. There are some simple steps that can be taken to help to cultivate self-awareness and deeply nourish the *self*:

- Making time for activities, experiences or therapeutic interventions that allow free movement of energy between body and mind, allowing thereby necessary attention to be paid to the expression of what lies in our heart and planning our self-growth, so that we can be open to new possibilities and adaptable to changing circumstances.

- Challenging ourselves to explore in a manner that is free of judgment, fear, others' demands, burning desire, blame, shame or anger who we truly are and our purpose, so that we can recognize the path that will allow us to fulfill our destiny.

- Keeping a check on whether our actions, thoughts and beliefs are in alignment so that we can live with a sense of authenticity in everything we do.

Practice no. 2: Sleeping soundly

Sleeping well is a critical step in achieving health in the microbiome. Most people are acutely aware of the extent to which sleep affects their health and wellbeing, and its significance in a Chinese Medicine diagnosis goes without saying. Acupuncturists habitually see a broad spectrum of sleep patterns and habits among their patients, from those who routinely prioritize sleep and enjoy good quality sleep every night, to those who are chronically sleep deprived, whether out of choice or not. Sleep and dreaming are considered important in Oriental Medicine, both diagnostically and therapeutically, due the connection with *Blood* and *Shen*, as well as the peaks and troughs of activities associated with each *Organ* across the night. Sleep deprivation can also speed up the rate at which *Jing* is depleted, thereby leading to premature aging and associated symptoms, including reduced cognitive functions and physical stamina, or lowered resistance to illness.

Sleep is similarly considered crucial within microbiome science, enabling the digestive system to rest fully and reset itself ahead of the new day. Furthermore, as already noted, sleep patterns affect appetite and satiety hormones, which can lead to inappropriate food choices and overeating, both of which may contribute to the development of conditions linked to microbiome dysbiosis. This association may at least in part explain the correlation that has been found between obesity and inadequate sleep (Cooper *et al.* 2018). Poor sleep also affects mood, cardiovascular functions, hormonal activity and breathing, all of which are connected to the microbiome (see Chapter 3) and therefore influence its health. This vicious cycle of sleep deprivation and an unhealthy microbial profile can further exacerbate health problems linked to poor sleep, including hypertension, anxiety, infertility and susceptibility to allergies.

Eastern and Western medicine's convergence on the importance of sleep is a clear signal of its therapeutic potential for all aspects of health and wellbeing, and also of the ways in which poor sleep may be tackled. The link between a healthy microbiome and sleep provides those suffering with insomnia or whose quality of sleep is suboptimal, a safe and natural way to address sleep issues, since there is a cross-referential interrelationship between sleep and the microbiome. In other words, sleep affects the microbiome and vice versa, and it is clear therefore that an integrated approach that takes into account and addresses different parameters of everyday health, whether sleep, diet or exercise, is the cornerstone of health, wellbeing and a thriving microbiome (see also Practice no. 3 below).

From an Oriental Medicine point of view, sleep epitomizes harmonious living with the natural cycles of day and night. Sleep is a *Yin* state of being, which mirrors the darkness of the night. Nourishing *Yin* provides the necessary balance to harness the *Yang* state of wakefulness during the day. The simple considerations of *Yin* and *Yang* can help patients understand appropriate sleep patterns and their importance from the perspective of both Oriental Medicine and the microbiome. Below are some simple pointers for healthy sleep, based on this *Yin–Yang* dynamic, which patients and practitioners may consider as part of a holistic approach to supporting the microbiome:

- When the days are longer (more *Yang*), less *Yin* (sleep and rest) is needed, and vice versa.

- The early light of spring and summer should be accompanied by an earlier rise.

- Creating a *Yin* sleeping environment—quiet, dark, still and cool—is important.

- Avoiding too much *Yang* before bedtime will ensure enough *Yin* is available to help induce sleep. This means not overeating in the evening or going to bed with a full stomach and being mindful that alcohol creates heat (*Yang*), which can disrupt sleep, especially for those who are already *Yin* deficient.

- Although water nourishes *Yin*, drinking too much water late in the day will inevitably impact on the *Water's Yang Organ*, the *Bladder*. This means avoiding drinking too much water in the evening so that there is less chance of being woken in the night with the need to go to the toilet.

- Understanding the quiet and nourishing stillness of the night helps fully realize the appropriateness of activities rather than create an overly restrictive regimen. Exercising, sport or the bright screens of computers and phones are clearly anathema to the mood of the night and as such belong to the *Yang* of the day.

- Preparations for the night should also ensure as much as possible that any emotional volatility or the active engagement of the mind is avoided. Here in particular, simple breathing exercises or meditative practices can help quieten the mind at bedtime.

- Connecting with the *self* and the microbiome may help too at bedtime. This active connection can be made by placing for a few moments a hot water bottle on the lower *Dantian* and *Mingmen Gate of Fire* (the area in the lower abdomen between the navel and pelvic bone and on the lower back just above the sacrum), as the warmth generates a deeply nourishing *Yin* energy in the area.

- Dietary recommendations may also be used to tackle any sleep problems.

Practitioners can further aid patients by considering individual diagnostic presentations and tailoring acupuncture treatments to help patients in their efforts to achieve good quality sleep, whether by nourishing *Blood* or *Yin* deficiency, or addressing their *Elemental* imbalance.

Practice no. 3: Eating well

It is important to remember, as noted earlier, that sleep, diet and the microbiome are linked, and that eating and sleeping patterns influence one another and the microbiome. This underlines the need for a holistic approach to nourishing the microbiome and how doing this can bring wide-ranging positive effects. As such, the principle of food as medicine, is, as already discussed in previous chapters, paramount.

Patients are often very receptive and keen to learn about nutrition and how it can support their acupuncture treatments. This provides practitioners of Oriental Medicine with an ideal opportunity to empower patients to take charge of their health and sustain treatment outcomes. For some patients, it may be that recommendations will be limited to some minor adjustments which draw on seasonal and other principles of food energetics (see Chapters 4 and 5).

For some patients, knowing how or where to start may seem challenging, especially if they are hesitant, resistant or simply disheartened and confused by the myriad, and perhaps seemingly unsuccessful, advice they have received or the strategies they have adopted over the years. There are also patients whose lifestyle or food choices seem to be at odds with what would in the view of a practitioner of Oriental Medicine represent a nourishing diet, even if the patients themselves may consider their diet as "healthy." Given that eating is a basic prerequisite to life, it is somewhat surprising that it has become such a complex subject and a source of frustration and even fear. How has something that should be so natural become so complicated? Of course, one of the reasons is that there is for many people an abundance of choices available, and with each of these choices comes a decision based on expectations, belief systems, convenience, taste, emotion, cultural norms, availability and price. From an evolutionary point of view, some of these choices are even more difficult to deal with, as humans' survival strategy has primed them to be as efficient as possible. This in part explains why foods with higher sugar and fat content are a combination that humans often find irresistible and even addictive: they provide a high level of energy in a short space of time. When such foods are consumed in small quantities—occasionally and at appropriate times, for instance during intense energy expenditure—they are not necessarily problematic. However, today, in societies where such foods are both imbued with symbolism—fun, childhood, reward, celebration, gifting and so on—and available on tap, they are often consumed in excess. Regular and excessive consumption of such treats can quickly become habitual and alter the microbial profile in a way that encourages more sugar to be consumed, turning what should be an occasional treat into a daily habit.

Breaking such cycles can be overcome with Oriental Medicine's *Microbiome Way* of eating, which has been discussed in earlier chapters, since it simply shows which foods should be prioritized and which ones should be limited, using pointers such as seasonality and balance of ingredients. As noted in Chapter 4, the fiber mantra and the inclusion of fermented foods are especially supportive of the microbiome. Broader principles should, however, be emphasized too, and these include "sustainability," "variety" and "simplicity." It is sometimes assumed that eating well is merely about eating certain foods from time to time, or eating one superfood, a wonder ingredient or even a cocktail of food supplements. The impact of such short-term approaches is still unclear, and while the individual health benefits of particular foods labeled as "superfoods" can be significant, they appear limited compared with the effect of a stable and rich microbiome. As such, the focus should be on nourishing the microbiome on an ongoing basis through a varied, balanced and heavily plant-based diet, as evidence from the field of microbiome nutrition research has found that this brings clear physical and mental health benefits. As many micro-organisms have only a transient effect, patients need to find those foods and meals that allow them to have continuity in their approach. This will be helped by eating a variety of microbiome-supportive foods, ensuring a good nutritional balance and a range of options across the seasons. Whether it is a mix of different vegetables or fruits that provides the basis of a dish to which whole grains and a little protein are added, there are many quick and easy ways to have diversity and equilibrium in our diet, and to create a simple yet nourishing meal. This approach means that there are no foods that are truly off-limit, only an order of prioritization. This denotes the simplicity that lies at the heart of the *Microbiome Way*, as there are just a few basic principles to follow:

- A balance of food types over the course of the day with around half of the diet made up of vegetables and fruits, a quarter of grain-based food and another quarter of dairy, meat, fish and eggs, with added fats, spices and condiments to enhance flavors and digestibility.

- As often as possible, eating natural and if possible organic foods, including both fresh and fermented. The more processed the food, the lower down the priority list it should go.

- Aligning meals to the body clock and seasons. This means eating breakfast, having a balanced lunch and a very light dinner early in the evening.

- Taking time to eat, to enjoy the food and to avoid distractions while eating so that the energetic benefits and flavors can be fully experienced and appreciated. If food is not enjoyed or is rushed, it may be harder to digest or create imbalances.

The *Microbiome Way* of eating, with its emphasis on living harmoniously with ourselves and our environment, is an uncomplicated, intuitive and effortless approach that is not driven by expectations. This is crucial to remember as there is a risk as always when discussing food and nutrition with some patients that we end up causing a sense of burden or failure. Practitioners should therefore stress to their patients how it is by harnessing the innate power of nature, just as acupuncture does, that the health benefits of this approach will be both spontaneous and easily maintained. In short, looking after our microbiome through food and nutrition is not a diet, it provides a harmonizing way of life from which many health and wellbeing gains will be made.

Practice no. 4: Synchronizing with natural rhythms

In good health, the harmonious functioning of the human body, including the microbiome, is quite remarkable, working like clock-work. There is an inherent intelligence in nature and the human body to adapt, respond and fully exploit the conditions of its natural habitat. Some patients may be unaware of the extent to which the natural rhythms of the day and night, and the seasons, influence not just health and vitality but the microbiome too. The microbiome is directly and indirectly affected by circadian physiological functions and the daily rhythms of rest, activity and feeding (Voigt *et al.* 2016). However, there are, as noted in previous chapters, bidirectional communication pathways between key sites such as the gut microbiome and brain, meaning that problems associated with natural adaptions to night and day and metabolic functions may be linked to microbial imbalances. The importance of living harmoniously with daily rhythms has already been noted in relation to practices of sleep and eating. However, there are many other aspects that practitioners and patients may want to consider, both for treatment planning and for self-care recommendations.

Both Chinese Medicine's Body Clock, which divides each day into 12 two-hour phases marking an *Organ's* peak and lowest point of activity, and the *Elemental* associations of the seasons (see Appendices A and B) provide a well-defined strategy to create an appropriate energetic balance for the microbiome. Inevitably, patients' harmonization with each moment of the day may be beyond their control, but there will be instances when behavioral adaptations can be made and thereby open new avenues to nourishing the microbial balance. Each practitioner will have individual ways of interpreting these adaptations for their patients to define relevant suggestions. For instance, patients may notice a worsening of symptoms at a particular time of day, or there may be, as with "sundowning" among dementia sufferers, agitation and anxiety in the late afternoon or early evening around the time when *Kidney* (which controls fear) is at its peak. In such cases, patients and practitioners can work closely together to find ways to synchronize the patient and the *Organs'* functions through acupuncture treatments and recommendations for everyday practices.

Spending as much time as possible outdoors, particularly in natural environments that are imbued with the *Elemental* qualities of each season, is also a simple practice to create a harmonious energetic balance and attune to natural rhythms. Many cultural traditions seek to enhance these connections by celebrating specific times of year, such as harvest time, or through practices such as Japan's *Shinrin Yoku* (forest bathing). This connection with nature positively influences the diversity of micro-organisms, as it has for a long time been observed that children growing up in cities have a narrower microbial diversity than those living in rural settings. It may be that this increased diversity comes not only from exposure to a diverse range of naturally occurring bacteria and fungi which will be in the air and soil, but also because of the energetic stimulus enabled by connecting with nature in a meaningful way.

Living in synchrony may therefore be supported by:

- Being mindful of appropriate activities for every time of the day and night, using the principles of Chinese Medicine's Body Clock (see Appendix B) and paying attention to which times of day trigger specific imbalances.

- Living in harmony with the seasons, whether through our behavior, diets or energy expenditure.

- Consciously seeking to connect with the seasonal energetic shifts that occur and their manifestation in nature, including the changing light, sounds, temperatures, smells and the transformation of landscapes.

- Tuning into the emotions triggered by seasonal changes and their transformative power.

- Spending as much time as possible in nature or partaking in activities that reconnect us with the natural world and its transformational qualities.

Practice no. 5: Conscious living

A theme running through each of the everyday practices presented here is a basic understanding of the continuum between humans and nature, of which the microbiome's ecological system is part. Acupuncture theory, it was pointed out at the start of our journey, divides the body into areas that reflect the natural world, with names of acupuncture points that refer to valleys, marshes and mountains, each of which symbolizes specific energetic potentiality. Such analogies are not coincidental. Rather they demonstrate the shared ancestry and existence of all living things. Conceptually, the idea of a natural habitat within humans that echoes and works with the outside world provides us with important lessons on nourishing the balance of the microbiome through the simple virtue that, much like a harvest, rewards are reaped from a caring, respectful and loving relationship with ourselves. This conception is profound in its transformative power and yet at the same time provides the means to facilitate patients' engagement in their own care and healing. It brings to the fore a straightforward yet fundamental realization that "what is given" physically, mentally or spiritually to body and mind will be "given back." The way in which this dynamic is played out can be seen in practice with patients who are overly critical of themselves in some way and whose self-directed criticisms and unachievable standards trigger negative energetic feedback, which manifests as anxiety, extreme timidity, eating disorders, pain, fainting, fatigue or musculoskeletal problems. Conversely, looking after body and mind through the practices outlined in this chapter can both positively contribute to patients' physical and mental health and help maintain it over time.

The analogy of humans as a microcosm of nature is helpful as we are all increasingly aware of the need to protect the environment we live in by making everyday lifestyle choices that minimize the negative effects of human activity, or that positively contribute to the health of our planet. Similarly, protecting the fauna and flora within can help guide patients towards "conscious living," which is good for their own health, microbial balance and the environment. For example, choosing foods that are organic and sustainably produced can support individual health and both ethical and environmentally friendly farming practices. "Conscious living" can and should take place deep within ourselves too. Many people are increasingly conscious of the environmental impact of their behavior, and adopting a similar approach to looking after ourselves is therefore both timely and relevant to wider lifestyle choices. From being conscious of the physical and mental trade-offs of seeking perfection when using toxins and chemicals for aesthetic purposes, to understanding how to truly keep a healthy "inner terrain," humans and nature are one and the same, and our practices should reflect this. Just like a gardener must weed to keep crops healthy and abundant, clearing out mentally and physically unwanted or harmful substances, or removing situations and memories that can choke us, provides us with the means to create a healthy environment to thrive in. Most people would be horrified at the idea of throwing rubbish into their own garden, yet this is how all too often we treat our bodies and minds.

It is important to stress here that Oriental Medicine's central tenet of *moderation* is crucial and that a more natural approach to living is a way, not a rule, as any rigidity in our approach could be equally detrimental. Being fanatical about "clean living" is unhelpful if it becomes a toxic obsession in itself or forces us to make irrational decisions about our own or others' health. The point is to become more aware of the choices we make and their impact, including for the microbiome, so that decisions are balanced and positively affect ourselves, others and the environment we live in. This awareness can help practitioners and patients move away from a judgmental stance, or from a perspective fueled by anger or fear, towards one that is much more open and fluid, and more introspective, which questions and provides a rationale for decisions, weighs up risks and benefits, and connects us seamlessly and appropriately with our natural habitat, the people around us and the wilderness within us. This forges a path to an authenticity that allows us to act with integrity and nurtures all aspects of mind, body and spirit, letting us recognize the role and importance of different healing modalities and systems of medicine, so that we can reconnect with and trust the abilities of our bodies and minds.

Conscious living means nurturing ourselves by:

- Respecting our bodies and having a healthy relationship with ourselves.

- Being aware of what we are putting into our bodies, both for ourselves and the environment, so that we can act with integrity.

- Being kind to ourselves: moderation is not just simple and easier to achieve, it is often best.

- Trusting the ability of bodies and minds to heal, to respond appropriately to new situations or adapt to new conditions.

- Being objective in our assessment of what may benefit our health and open to new ideas and knowledge across disciplines.

- Environmentally friendly living and, when possible, opting for organic and sustainable products.

CONCLUSION: A SHARED PATH

The *Microbiome Way* is an approach that offers a helpful tool for both practitioners and patients to understand how to improve health and wellbeing, and which can be meaningfully integrated within the practice and experience of Oriental Medicine. It aligns well with its principles, since practicing and experiencing therapies such as acupuncture requires a high level of self-awareness and asks both practitioners and patients to let go of preconceived notions and labels of disease and be open to the possibility of healing. In this sense, the *Microbiome Way* provides an opportunity for a shared journey in which Oriental Medicine diagnosis and microbiome self-care can improve health outcomes, sustain results and prevent ill health in the longer term. It is an approach that deepens the therapeutic relationship, allowing practitioners and patients to experience ongoing healing and energetic balance rather than using a ready-made formula. The essence of Oriental Medicine lies in its understanding that it is the person rather than the condition that is treated, and it is this very principle that underpins the *Microbiome Way* and the symbiotic relationships it affords.

Conclusion

It is hoped that as we reach the end of this book, patients and practitioners working across the many, both old and new, Eastern and Western, traditions of medicine will feel inspired and intrigued by the formidable potential for health that is within our reach when the extraordinary science of the microbiome is integrated into the profound and timeless teachings of Oriental Medicine. This integration is especially pertinent in a world in which we are fortunate enough to benefit from continual, often life-saving, medical and scientific progress, yet in which we have also gradually become distanced from the authenticity, spontaneity and innate ecological balance of the natural world from which we come, to which we belong and of which we are made. Traditional therapies such as those contained within Oriental Medicine help us return to that natural state of being and harness the healing potential that comes with it. For this reason, there is much to be gained from relearning the wisdom from older times (Dechar 2006) while embracing the added value of modern scientific discoveries, including cutting-edge microbiome research. The microbiome epitomizes this juncture between traditional and modern, and when examined through the lens of Oriental Medicine shows us the remarkable synergy of Eastern and Western medical knowhow, and how the all-too-often perceived incompatibility of these systems may be the exception rather than the rule. Far from being at odds, trailblazing microbiome research is in fact helping validate many assumptions and therapeutic stances taken in Oriental Medicine. The emphasis within Oriental Medicine that spontaneous healing occurs within set conditions mirrors the very mechanisms by which health in the microbiome is seen to influence and regulate many aspects of both physical and mental health. Significantly too, at the heart of these two conceptions of health are the patients: physical and spiritual beings whose existence is influenced and defined by their symbiotic relationships with the self and its microbial residents. This realization shows us that there are many ways to become a healthier, happier and stronger person, not least by following the *Elemental* principles that intertwine with parameters of microbial health, and which have been laid out in earlier chapters.

The correspondence between Oriental Medicine and the microbiome brings to the fore health variables that may have been given little consideration until now, and in particular how the experience and outcomes of therapies such as acupuncture may be influenced by microbiome health, or how microbial profiles may be altered through such therapies. Indeed, given the microbiome's ability to influence physical and mental functions, it is conceivable that it could also affect energetic receptivity to acupuncture—a point of significance for practitioners in their diagnostic and therapeutic approaches. The theoretical and empirical cross-overs between Oriental Medicine and microbiome science certainly suggest new pathways to recovery for patients, while for patients themselves, the lesson is a simple one. Wellness does not have to be complicated. More often than not, we forget how much we can do to create the right environment for health and wellbeing with nourishing lifestyle choices and relationships—both externally and within ourselves—so that we can experience everyday wellness, physically and mentally. Self-awareness and self-care strategies learnt from Eastern traditions (see Deadman 2016) are the foundations of the everyday wellness that is nurtured through therapeutic interventions such as acupuncture and are aided by our microbial universe within. It is from this nourishment to body and mind that trust in the process of life, self-belief, integrity and resilience can grow, allowing healing to occur and good health to prevail. Tangible as the potential afforded by this complementarity between Eastern and Western medicine might be, finding and following our own *Microbiome Way* may require considerable changes to be made, including to our understanding of health, to how we live, eat, breathe, sleep and more generally look after ourselves. As hard as change can be, it is also an opportunity that is both essential to progress and intrinsic to all living things. Change is, after all, what life is all about, and our adaptability and willingness to change certainly is key when it comes to good health.

Elemental Guide to Selected Microbiome Beneficial Foods

B elow are some examples of foods organized around their *Elemental* affinity and potential harmonizing effects on each of the *Elements*, when consumed in moderation. These categories should not be seen as mutually exclusive, and instead should be used to inform a balanced approach that ensures all *Organs* are supported through dietary choices that also enable seasonal foods to be used when possible. Organic varieties are preferred as these are seen as offering ecological benefits for both humans and their environment.

EARTH

Effects: Seen as providing a grounding energetic effect through their natural slightly sweet flavor, *Earth* energy-enhancing foods are especially nourishing to *Spleen* and *Stomach* and are gentle on the digestive system as a whole. Due to this role, these are foods that can be included at all times, although the peak time for seasonal *Earth* foods is late summer. The *Earth Organs* are most receptive in the morning (7am–11am), and eating a well-balanced and nutritive breakfast provides, therefore, the foundation of a well-nourished body and mind.

Cooking methods: Steamed, boiled, stewed.

Supportive fiber-rich foods: Carrots, zucchinis/courgettes, dates, apples, plums, pumpkin, butternut squash, turnip, swede, peas, sweet potatoes, potatoes, chickpeas, yellow soya beans, apricots, brown rice, oats, pumpkin seeds, flaxseed, pine kernels, chestnuts, hazelnuts, peanuts.

Spices and herbs: Cumin, marjoram, nutmeg, turmeric.

Fermented foods/probiotics: Hard and soft cheeses, such as Cheddar, Gruyère, Gouda, Brie, Parmesan.

Suitable fats and condiments: Sunflower oil, pumpkin seed oil, honey.

Other proteins: Anchovy, mullet, tuna, salmon, monkfish, chicken, turkey.

Drinks: Rosehip tea, licorice tea, Ginseng tea, rice wine.

METAL

Effects: The energetic effect of the *Metal Element* is purifying and clearing, providing a sense of lightness and balance in both body and mind. By having a decongestant energetic effect through their pungent flavors, these foods can also help resolve *Dampness* and may therefore help address patterns of dysbiosis. Due to the *Elemental* associations, seasonal autumnal foods as well as many lighter-colored foods will bring beneficial energetic effects to its corresponding *Organs—Lung* and *Large Intestine*. In addition to having direct relevance to the microbiome through the functions of the *Large Intestine* and through the mechanisms of the gut–lung axis, the *Metal Element* plays an important role in the circulation of *Qi*, which is necessary for all functions across body systems, including protective, restorative and regulatory tasks.

Cooking methods: Stir-fried, baked, boiled.

Supportive fiber-rich foods: Bran, rice, blueberries, plums, mushrooms, garlic, onions, leeks, parsnips, pears, fennel, cabbage, rhubarb, almonds, cashews, peanuts, white sesame seeds.

Spices and herbs: Cinnamon, fenugreek seeds, chives, cardamom, basil, sage, tarragon.

Fermented foods/probiotics: Cabbage sauerkraut, fennel sauerkraut.

Suitable fats and condiments: Peanut oil, mustard, wasabi.

Other proteins: Herring, chicken eggs, cod, pollock, tofu.

Drinks: Ginger tea, jasmine tea, white tea.

WATER

Effects: Associated with winter, the most *Yin* time of the year, the *Water Element* has a very cooling and calming effect, which allows for strength and power to be stored and called upon when needed. The associated *Organs*—*Kidney* and *Bladder*—reflect this energetic and can quickly become depleted. For optimal health, these *Organs* remind us that it is important to ensure adequate hydration, salt balance and sufficient periods of rest to recharge. Special care must be taken at this time of year to avoid excessively cooling foods, focusing instead on bringing warmth to balance the cooler temperatures, preserving energy as much as possible and on ensuring body and mind are not being put under unnecessary strain. With its connection to *Jing* and *Mingmen Dantian*, supporting the *Water Element* may strengthen health in the microbiome and in particular the re-establishment of symbiotic relationships.

Cooking methods: Fried, roasted, grilled.

Supportive fiber-rich foods: Kale, Swiss chard, broad beans, beetroot, red cabbage, eggplant/aubergine, seaweed (kelp), celery, Brussels sprouts, potatoes, blackberries, cranberries, adzuki beans, kidney beans, black-eyed peas, walnuts, black sesame seeds, chestnuts, poppy seeds, chia, barley, millet, wild rice, (whole)wheat noodles/pasta.

Spices and herbs: Parsley, caraway, clove, sage, thyme.

Fermented foods/probiotics: Tempeh, Roquefort, Stilton, kombucha, kimchi.

Suitable fats and condiments: Miso, sesame oil, soy sauce, walnut oil.

Other proteins: Lamb, halibut, bream.

Drinks: Raspberry leaf tea, parsley tea, red wine.

WOOD

Effects: Providing an upward energy necessary for growth, movement and flexibility, the *Wood Element* and its associated *Organs—Liver* and *Gallbladder*—are at their peak during spring and benefit from eating fresh seasonal foods, particularly green-colored vegetables and those with a sour flavor that gently cleanses and detoxifies mind and body. The energy is also said to be aided by forms of exercise such as *Tai Chi*, *Qigong*, *Yoga* and walking, as they help clear a person's system of stagnant energy, including *Dampness*. As such, supporting health in *Liver* and *Gallbladder* is highly beneficial to help regulate balance in the microbiome.

Cooking methods: Steamed, boiled, stir-fried.

Supportive fiber-rich foods: Quinoa, wheatberries, broccoli, spring greens, spinach, pak-choi, choi-sum, green banana, celeriac, radishes, watercress, new potatoes, asparagus, broad beans, lemons, black soybean, Brazil nuts, hemp seeds, fennel seeds, olives.

Spices and herbs: Coriander, mint.

Fermented foods/probiotics: Kefir, yoghurt, sourdough.

Suitable fats and condiments: Olive oil, butter, raw vinegar.

Other proteins: Beef, mackerel, trout, sea bass, plaice.

Drinks: Green tea, nettle tea, black tea, peppermint tea.

FIRE

Effects: As the *Element* associated with the summer months, supporting this energy through food and self-care provides warmth, vitality and a sense of being connected with ourselves and those around us, thereby imparting an ability to feel animated, spirited and passionate about life, relationships and work. Vibrant colors dominate the fruits and vegetables available at this time of year and can be enjoyed alongside foods that help balance the rising summer temperatures. Enjoyment of food is itself nourishing to the *Fire* energy, and sharing a meal with friends and family or eating in the open air to connect with the lightness of summer can provide additional qualities and strength to the *Element*. There are four *Organs* within the *Fire Element—Heart, Small Intestine, San Jiao* and *Pericardium*—and they each have major regulatory and nutritive functions, both physically and mentally. The *Fire Element* can be seen to have a direct influence on the microbiome, and vice versa, through the gut–brain axis. Food therapeutics that ensure health in this *Element* may therefore provide a useful approach to harmonizing health in both body and mind, including in the microbiome.

Cooking methods: Raw, steamed.

Supportive fiber-rich foods: Raw oats, wheatgerm, rye, spelt, bulgur wheat, avocados, tomatoes, peppers, red lentils, watermelon, cantaloupe melon, cherries, apricots, red grapes, strawberries, raspberries, sunflower seeds, pistachios, dark chocolate.

Spices and herbs: Rosemary, basil, oregano, mint, cayenne, chili, black pepper.

Fermented foods/probiotics: Goat cheese, feta, coconut yoghurt.

Suitable fats and condiments: Coconut oil, grape seed oil.

Other proteins: Sole, chicken eggs, duck eggs.

Drinks: Lime flower tea, hibiscus tea, coffee, light wines.

Chinese Medicine's Body Clock

A simple body clock system is used in classical acupuncture to understand and treat according to the peaks and troughs of the *Organs'* activities across different times of day and night. According to Chinese Medicine's Body Clock, each 24-hour cycle is divided into 12 two-hour periods. During each two-hour period, one of the 12 *Organs* from the five *Elements* dominates, thus taking responsibility in turn for proper daily functioning of mind and body at specific times.

WOOD ELEMENT

- 11pm–1am *Gallbladder*: The *Organ* is said to help with decision-making, allowing thoughts from the day to be organized and contemplated quietly and at subconscious level while we sleep.

- 1am–3am *Liver*: With its direct connection to *Blood*, *Liver* hours are a time when we should feel deeply anchored in our sleep to allow detoxification of both body and mind.

METAL ELEMENT

- 3am–5am *Lung*: This is seen in many Eastern practices as an optimal time to enter a deep state of meditation and it is still today the time when monks in Eastern Asia rise for this purpose. While very early risers who also go to bed very early may find it a good time to engage in tasks that demand crystal-clear thinking, for most people this is, however, a good time to still be asleep and to be resting, as this will strengthen breathing and immunity.

- 5am–7am *Large Intestine*: A time to get up. Starting the day with a glass of water can help cleanse the body and having a bowel movement at this time will also help the body clear unwanted waste and be more receptive to a nourishing breakfast.

EARTH ELEMENT

- 7am–9am *Stomach*: The ideal time to have a truly satisfying breakfast, which should provide a good nutritional basis for the day.

- 9am–11am *Spleen*: Associated with the power of concentration, this can be a good time to study or to draw on the power of our mental faculties.

FIRE ELEMENT

- 11am–1pm *Heart*: Having lunch around this time can help balance energetic shifts as we reach the most *Yang* time of the day. Taking time also to step out of the office or home to get sunlight will help harmonize our internal body clock with natural cycles and rhythms.

- 1pm–3pm *Small Intestine*: The *Organ* which sorts the pure from the impure, the digestive system will be working on distributing and absorbing nutrients as needed. It can also be a good time to focus on tasks that involve sorting and sifting.

WATER ELEMENT

- 3pm–5pm *Bladder*: As a *Water Organ*, hydration is critical during these hours. The *Organ* epitomizes skill, immense power and resilience, and demanding tasks will more easily be achieved at this time.

- 5pm–7pm *Kidney*: Although this is a time when many are still in their office or are heading to the gym, this can in fact be a good time for activities that nourish our *Yin* energy, including gentle forms of exercise such as *Yoga*, walking or *Qigong*, or eating a light dinner that is calming and gentle on our digestive system.

FIRE ELEMENT

- 7pm–9pm *Pericardium*: Part of the *Fire Organs*, the *Pericardium* hours are well suited to making connections with those around us, whether spending time with family, seeing friends, socializing with neighbors or others with common interests and hobbies, or doing that all-important personal or professional networking.

- 9pm–11pm *San Jiao*: The *Organ* associated with homeostatic functions and balance; this is a good time to make sure that we keep everything in moderation as we get ready for the night. Whether it be taking a warm bath or shower, reading or listening to music with a low light, activities that help calm and regulate body systems should be prioritized.

References

Abbott, A. (2016) 'Scientists bust myth that our bodies have more bacteria than human cells.' *Nature.* Accessed on 22/08/2021 at www.nature.com/articles/nature.2016.19136

Amon, P. and Sanderson, I. (2017) 'What is the microbiome?' *Archives of Disease in Childhood: Education & Practice 102*, 5, 257–260. Accessed on 18/08/2021 at https://ep.bmj.com/content/102/5/257.long

Atkins, P. (2013) *Facial Enhancement Acupuncture: Clinical Use and Application.* London: Singing Dragon.

Baker, M. (2016) *A Chronological Journey through Chinese Medical History on the Causes of Disease.* Bath: Brown Dog.

Belizário, J.E. and Faintuch, J. (2018) 'Microbiome and Gut Dysbiosis.' In R. Silvestre and E. Torrado (eds) *Metabolic Interaction in Infection. Experientia Supplementum.* Cham: Springer.

Berg, G., Rybakova, D., Fischer, D., Cernava, T. *et al.* (2020) 'Microbiome definition re-visited: Old concepts and new challenges.' *Microbiome 8*, 103. Accessed on 18/08/2021 at https://microbiomejournal.biomedcentral.com/track/pdf/10.1186/s40168-020-00875-0.pdf

Bondar, T. (2019) 'Regulation of mucosal immunity by the microbiota.' *Nature Portfolio.* Accessed on 19/08/2021 at www.nature.com/articles/d42859-019-00014-2

Bray, N. (2019) 'The microbiota–gut–brain axis.' *Nature Portfolio.* Accessed on 19/08/2021 at www.nature.com/articles/d42859-019-00021-3

Breit, S., Kupferberg, A., Rogler, G. and Gregor, H. (2018) 'Vagus nerve as modulator of the brain–gut axis in psychiatric and inflammatory disorders.' *Frontiers in Psychiatry 9.* Accessed on 22/08/2021 at www.frontiersin.org/articles/10.3389/fpsyt.2018.00044/full

Butler, M.I., Mörkl, S., Sandhu, K.V., Cryan, J. and Dinan, T.G. (2019) 'The gut microbiome and mental health: What should we tell our patients?: Le microbiote intestinal et la santé mentale : Que devrions-nous dire à nos patients?' *The Canadian Journal of Psychiatry 64*, 11, 747–760. Accessed on 19/08/2021 at https://journals.sagepub.com/doi/full/10.1177/0706743719874168

Byrd, A.L., Belkaid, Y. and Segre, J.A. (2018) 'The human skin microbiome.' *Nature Reviews Microbiology 16*, 143–155. Accessed on 21/08/2021 at www.nature.com/articles/nrmicro.2017.157

Campbell, L. (2019) *Ten Powerful Steps to Clear Psoriasis*. Amazon Books.

Carabotti, M., Scirocco, A., Maselli, M.A. and Severi, C. (2015) 'The gut–brain axis: Interactions between enteric microbiota, central and enteric nervous systems.' *Annals of Gastroenterology 28*, 2, 203–209. Accessed on 22/08/2021 at www.ncbi.nlm.nih.gov/pmc/articles/PMC4367209

Carpenter, S. (2012) 'That gut feeling.' *Monitor on Psychology 43*, 8. Accessed on 21/08/2021 at www.apa.org/monitor/2012/09/gut-feeling

Castillo, D.J., Rifkin, R.F., Cowan, D.A. and Potgieter, M. (2019) 'The healthy human blood microbiome: Fact or fiction?' *Frontiers in Cellular Infectious Microbiology 9*, 148. Accessed on 22/08/2021 at www.frontiersin.org/articles/10.3389/fcimb.2019.00148/full

Chial, H. and Craig, J. (2008) 'mtDNA and mitochondrial diseases.' *Nature Education 1*, 1, 217. Accessed on 18/08/2021 at www.nature.com/scitable/topicpage/mtdna-and-mitochondrial-diseases-903

Clark, A. and Mach, N. (2016) 'Exercise-induced stress behavior, gut–microbiota-brain axis and diet: A systematic review for athletes.' *Journal of the International Society of Sports Nutrition 13*, 43. Accessed on 22/08/2021 at https://jissn.biomedcentral.com/articles/10.1186/s12970-016-0155-6

Clutter, C. (2019) 'Disappearance of the human microbiota: How we may be losing our oldest allies.' American Society for Microbiology. Accessed on 18/08/2021 at https://asm.org/Articles/2019/November/Disappearance-of-the-Gut-Microbiota-How-We-May-Be

Collen, A. (2015) *10% Human: How Your Body's Microbes Hold the Key to Health and Happiness*. London: William Collins.

Cooper, C.B., Neufeld, E.V., Dolezal, B.A. and Martin, J.L. (2018) 'Sleep deprivation and obesity in adults: A brief narrative review.' *BMJ Open Sport & Exercise Medicine 4*, 1. Accessed on 22/08/2021 at www.ncbi.nlm.nih.gov/pmc/articles/PMC6196958

Cummings, J. and Engineer, A. (2018) 'Denis Burkitt and the origins of the dietary fiber hypothesis.' *Nutrition Research Reviews 31*, 1, 1–15. Accessed on 21/08/2021 at www.cambridge.org/core/journals/nutrition-research-reviews/article/denis-burkitt-and-the-origins-of-the-dietary-fibre-hypothesis/1DA569CF06DB93A4FF2DA54629A5D566

Davenport, E.R., Mizrahi-Man, O., Michelini, K., Barreiro, L.B., Ober, C. and Gilad, Y. (2014) 'Seasonal variation in human gut microbiome composition.' *Plos One 9*, 3. Accessed on 19/08/2021 at https://journals.plos.org/plosone/article?id=10.1371/journal.pone.0090731

Deadman, P. (2016) *Live Well, Live Long: Teachings from the Chinese Nourishment of Life Tradition*. Hove: Journal of Chinese Medicine.

Dechar, L.E. (2006) *Five Spirits*. New York: Lantern Books.

Dekaboruah, E., Suryavanshi, M., Chettri, D. and Verma, A. (2020) 'Human microbiome: An academic update on human body site specific surveillance and its possible role.' *Archives of Microbiology 202*, 2147–2167. Accessed on 22/08/2021 at https://link.springer.com/article/10.1007/s00203-020-01931-x

den Besten, G., van Eunen, K., Groen, A.K., Venema, K., Reijngood, D.J. and Bakker, B.M. (2013) 'The role of short-chain fatty acids in the interplay between diet, gut microbiota, and host energy metabolism.' *Journal of Lipid Research 54*, 9, 2325–2340. Accessed on 22/08/2021 at www.jlr.org/article/S0022-2275(20)35124-5/fulltext

Dimidi, E., Cox, S.R., Rossi, M. and Whelan, K. (2019) 'Fermented foods: Definitions and characteristics, impact on the gut microbiota and effects on gastrointestinal health and disease.' *Nutrients 11*, 8, 1806. Accessed on 22/08/2021 at www.mdpi.com/2072-6643/11/8/1806/htm

Durack, J. and Lynch, S.V. (2019) 'The gut microbiome: Relationships with disease and opportunities for therapy.' *Journal of Experimental Medicine 216*, 1, 20–40. Accessed on 18/08/2021 at www.ncbi.nlm.nih.gov/pmc/articles/PMC6314516

Eck, A., Rutten, N., Singendonk, M., Rijkers, G. *et al.* (2020) 'Neonatal microbiota development and the effect of early life antibiotics are determined by two distinct settler types.' *PLoS ONE 15*, 2. Accessed on 18/08/2021 at https://journals.plos.org/plosone/article?id=10.1371/journal.pone.0228133

Eisenstein, M. (2020a) 'The skin microbiome.' *Nature 588*, S209. Accessed on 18/08/2021 at www.nature.com/articles/d41586-020-03523-7

Eisenstein, M. (2020b) 'The hunt for a healthy microbiome.' *Nature 577*. Accessed on 18/08/2021 at https://media.nature.com/original/magazine-assets/d41586-020-00193-3/d41586-020-00193-3.pdf

Enaud, R., Prevel, R., Ciarlo, E., Beaufils, F. *et al.* (2020) 'The gut–lung axis in health and respiratory diseases: A place for inter-organ and inter-kingdom crosstalks.' *Frontiers in Cellular and Infection Microbiology 10*, 9. Accessed on 21/08/2021 at www.frontiersin.org/article/10.3389/fcimb.2020.00009

Enders, G. (2015) *Gut*. London: Scribe Publications.

European Society for Neurogastroenterology and Motility (2021) 'Gut microbiota info.' Accessed on 18/08/2021 at www.gutmicrobiotaforhealth.com/about-gut-microbiota-info

Farrell, A. (2019) 'The importance of feeding your microbiota.' *Nature Portfolio*. Accessed on 18/08/2021 at www.nature.com/articles/d42859-019-00015-1

Farré-Maduell, E. and Casals-Pascual, C. (2019) 'The origins of gut microbiome research in Europe: From Escherich to Nissle.' *Human Microbiome Journal 14*. Accessed on 18/08/2021 at www.sciencedirect.com/science/article/pii/S2452231719300144

Fehervari, Z. (2019) 'Mechanisms of colonization resistance.' *Nature Portfolio*. Accessed on 18/08/2021 at www.nature.com/articles/d42859-019-00018-y

Firth, J., Gangwisch, J.E., Borsini, A., Wootton, R.E. and Mayer, E.A. (2020) 'Food and mood: how do diet and nutrition affect mental wellbeing?' *British Medical Journal 369*. doi:10.1136/bmj.m2382

Foster, J.A., Rinaman, L. and Cryan, J.F. (2017) 'Stress and the gut–brain axis: Regulation by the microbiome.' *Neurobiology of Stress 7*, 124–136. Accessed on 18/08/2021 at www.sciencedirect.com/science/article/pii/S2352289516300509

Franglen, N. (2014) *The Handbook of Five Element Practice*. London: Singing Dragon.

Fukui, H. (2019) 'Role of gut dysbiosis in liver diseases: What have we learned so far?' *Diseases 7*, 4, 58. Accessed on 22/08/2021 at www.ncbi.nlm.nih.gov/pmc/articles/PMC6956030

Ganguly, P. (2019) 'Microbes in us and their role in human health and disease.' National Human Genome Research Institute. Accessed on 18/08/2021 at www.genome.gov/news/news-release/Microbes-in-us-and-their-role-in-human-health-and-disease

García, H. and Miralles, F. (2016) *Ikigai: The Japanese Secret to a Long and Happy Life*. London: Hutchinson.

Harvard T.H. Chan School of Public Health (2021a) 'The microbiome.' Accessed on 19/08/2021 at www.hsph.harvard.edu/nutritionsource/microbiome

Harvard T.H. Chan School of Public Health (2021b) 'The nutrition source: Whole grains.' Accessed on 22/08/2021 at www.hsph.harvard.edu/nutritionsource/what-should-you-eat/whole-grains

Hofer, U. (2019) 'Microbiome analyses in large human populations.' *Nature Portfolio*. Accessed on 18/08/2021 at www.nature.com/articles/d42859-019-00020-4

Jang, J.-H., Yeom, M.-J., Ahn, S., Oh, J.-Y. *et al.* (2020) 'Acupuncture inhibits neuroinflammation and gut microbial dysbiosis in a mouse model of Parkinson's disease.' *Brain, Behavior, and Immunity 89*, 641–655. Accessed on 22/08/2021 at www.sciencedirect.com/science/article/pii/S0889159120307285?via%3Dihub

Jarrett, L. (2006) *The Clinical Practice of Chinese Medicine*. Stockbridge, MA: Spirit Path.

Johnson, K.V.A. (2020) 'Gut microbiome composition and diversity are related to human personality traits.' *Human Microbiome Journal 15*. Accessed on 22/08/2021 at www.sciencedirect.com/science/article/pii/S2452231719300181

Kaptchuk, T.J. (2000) *Chinese Medicine*. London: Rider.

Katagiri, S., Shiba, T., Tohara, H., Yamaguchi, K. *et al.* (2019) 'Re-initiation of oral food intake following enteral nutrition alters oral and gut microbiota communities.' *Frontiers in Cellular Infectious Microbiology 9*, 434. Accessed on 22/08/2021 at www.frontiersin.org/articles/10.3389/fcimb.2019.00434/full

Katz, S.E. (2012) *The Art of Fermentation*. White River Junction, VT: Chelsea Green.

Keown, D. (2014) *The Spark in the Machine*. London: Singing Dragon.

Kim T.W., Jeong, J.-H. and Hong, S.-C. (2015) 'The impact of sleep and circadian disturbance on hormones and metabolism.' *International Journal of Endocrinology.* Accessed on 18/08/2021 at www.ncbi.nlm.nih.gov/pmc/articles/PMC4377487

Larre, C. and de la Vallée, E. (1995) *Rooted in Spirit.* New York: Station Hill Press.

Le Roy, C., Wells, P.M., Si, J., Raes, J., Bell, J.T. and Spector, T.D. (2020) 'Red wine consumption associated with increased gut microbiota α-diversity in 3 independent cohorts.' *Gastroenterology 158*, 270–272. Accessed on 22/08/2021 at www.researchgate.net/publication/335446113_Red_Wine_Consumption_Associated_With_Increased_Gut_Microbiota_a-Diversity_in_3_Independent_Cohorts/link/5dfc9a7692851c83648b7d38/download

Lederer, A.-K., Pisarski, P., Kousoulas, L., Fichtner-Feigl, S., Hess, C. and Huber, R. (2017) 'Postoperative changes of the microbiome: Are surgical complications related to the gut flora? A systematic review.' *BMC Surgery 17*, 1, 125. Accessed on 18/08/2021 at www.ncbi.nlm.nih.gov/pmc/articles/PMC5715992

Lee, Y.B., Byun, E.J. and Kim, H.S. (2019) 'Potential role of the microbiome in acne: A comprehensive review.' *Journal of Clinical Medicine 8*, 7, 987. Accessed on 22/08/2021 at www.mdpi.com/2077-0383/8/7/987/htm

Leggett, D. (2014a) *Recipes for Self-Healing.* Totnes: Meridian Press.

Leggett, D. (2014b) *Helping Ourselves.* Totnes: Meridian Press.

Liang, X. and FitzGerald, G.A. (2017) 'Timing the microbes: The circadian rhythm of the gut microbiome.' *Journal of Biological Rhythms 32*, 505–515. Accessed on 21/08/2021 at https://journals.sagepub.com/doi/10.1177/0748730417729066

Lo, B.C., Chen, G.Y., Núñez, G. and Caruso, R. (2021) 'Gut microbiota and systemic immunity in health and disease.' *International Immunology 33*, 4, 197–209. Accessed on 22/08/2021 at https://academic.oup.com/intimm/article/33/4/197/6047278

Lozupone, C.A., Stombaugh, J.I., Gordon, J.I., Jansson, J.K. and Knight, R. (2012) 'Diversity, stability and resilience of the human gut microbiota.' *Nature 489*, 220–230. Accessed on 21/08/2021 at www.nature.com/articles/nature11550

Lyon, L. (2018) '"All disease begins in the gut": Was Hippocrates right?' *Brain 141*, 3, e20. Accessed on 26/10/2021 at https://academic.oup.com/brain/article/141/3/e20/4850980

Maciocia, G. (1998) *The Foundations of Chinese Medicine.* Edinburgh: Churchill Livingstone.

Makki, K., Deehan, E.C., Walter, J. and Bäckhed, F. (2018) 'The impact of dietary fiber on gut microbiota in host health and disease.' *Cell Host and Microbe 23*, 6, 705–715. Accessed on 26/10/2021 at www.sciencedirect.com/science/article/pii/S193131281830266X

Manach, C., Scalbert, A. Morand, C., Rémésy, C. and Jiménez, L. (2004) 'Polyphenols: Food sources and bioavailability.' *The American Journal of Clinical Nutrition*

79, 5, 727–747. Accessed on 23/08/2021 at https://academic.oup.com/ajcn/article/79/5/727/4690182

Marchesi, J., Adams, D., Fava, F., Hermes, G. *et al.* (2016) 'The gut microbiota and host health: A new clinical frontier.' *Gut 65*, 2, 330–339. Accessed on 23/08/2021 at https://gut.bmj.com/content/65/2/330.long

Meyer, A., Laverny, G., Bernardi, L., Charles, A.L. *et al.* (2018) 'Mitochondria: An organelle of bacterial origin controlling inflammation frontiers.' *Immunology 9*, 536. Accessed on 18/08/2021 at www.frontiersin.org/article/10.3389/fimmu.2018.00536

Milani, C., Duranti, S., Bottacini, F., Casey, E. *et al.* (2017) 'The first microbial colonizers of the human gut: Composition, activities, and health implications of the infant gut microbiota.' *Microbiology and Molecular Biology Reviews 81*, 4. Accessed on 22/08/2021 at www.ncbi.nlm.nih.gov/pmc/articles/PMC 5706746

Miller, I. (2018) 'The gut–brain axis: Historical reflections.' *Microbial Ecology in Health and Disease 29*, 2. Accessed on 22/08/2021 at www.tandfonline.com/doi/full/10.1080/16512235.2018.1542921?scroll=top&needAccess=true

Mitchell, S. (transl.) (1999) *Tao Te Ching: An Illustrated Journey.* Translation of Lao Tzu—Tao Te Ching (The Book of the Way). London: Frances Lincoln.

Morrison, D.J. and Preston, T. (2016) 'Formation of short chain fatty acids by the gut microbiota and their impact on human metabolism.' *Gut Microbes 7*, 3, 189–200. Accessed on 18/08/2021 at www.tandfonline.com/doi/full/10.1080/19490976.2 015.1134082

Mosley, M. (2017) *The Clever Guts Diet.* London: Shortbooks.

Nardone, G. and Compare, D. (2015) 'The human gastric microbiota: Is it time to rethink the pathogenesis of stomach diseases?' *United European Gastroenterology Journal 3*, 3, 255–260. Accessed on 21/08/2021 at www.ncbi.nlm.nih.gov/pmc/articles/PMC4480535

Nelson-Dooley, C. (2019) *Heal Your Oral Microbiome.* New York: Ulysses Press.

New Scientist (2016) 'How lack of oxygen makes bacteria cause acne and how to stop it.' 28 October. Accessed on 22/08/2021 at www.newscientist.com/article/2110826-how-lack-of-oxygen-makes-bacteria-cause-acne-and-how-to-stop-it

New York Times (2018) 'Does the gut ever fully recover from antibiotics?' 21 December. Accessed on 22/08/2021 at www.nytimes.com/2018/12/21/well/live/does-the-gut-microbiome-ever-fully-recover-from-antibiotics.html

Ni, M. (1995) *The Yellow Emperor's Classic of Medicine: A New Translation of the Neijing Suwen.* Boston, MA: Shamhala.

Ochoa-Repáraz, J. and Kasper, L.H. (2016) 'The second brain: Is the gut microbiota a link between obesity and central nervous system disorders?' *Current Obesity Reports 5*, 1, 51–64. Accessed on 18/08/2021 at www.ncbi.nlm.nih.gov/pmc/articles/PMC4798912

Otto, M. (2017) '*Staphylococcus epidermidis*: A major player in bacterial sepsis?' *Future Microbiology 12*, 12, 1031–1033. Accessed on 20/10/2021 at https://doi.org/10.2217/fmb-2017-0143

Paoli, A., Tinsley, G., Bianco, A. and Moro, T. (2019) 'The influence of meal frequency and timing on health in humans: The role of fasting.' *Nutrients 11*, 4, 719. Accessed on 22/08/2021 at www.mdpi.com/2072-6643/11/4/719/htm

Patrimoine Culinaire Suisse (2020) *Birchermus*. Accessed on 22/08/2021 at www.patrimoineculinaire.ch/Produit/Birchermus-Bircher/175

Pennisi, E. (2020) 'Meet the psychobiome: The gut microbiome that may alter how you think, feel and act.' *Science*. Accessed on 19/08/2021 at www.sciencemag.org/news/2020/05/meet-psychobiome-gut-bacteria-may-alter-how-you-think-feel-and-act

Pitchford, P. (2002) *Healing with Whole Foods*. Berkeley, CA: North Atlantic Books.

Qi, Q., Liu, Y.N., Jin, X.M., Zhang, L.S. *et al.* (2018) 'Moxibustion treatment modulates the gut microbiota and immune function in a dextran sulphate sodium-induced colitis rat model.' *World Journal of Gastroenterology 24*, 28, 3130–3144. Accessed on 22/08/2021 at www.ncbi.nlm.nih.gov/pmc/articles/PMC6064969

Qi, X., Yun, C., Pang, Y. and Qiao, J. (2021) 'The impact of the gut microbiota on the reproductive and metabolic endocrine system.' *Gut Microbes 13*, 1, 1–21. Accessed on 19/08/2021 at www.tandfonline.com/doi/full/10.1080/19490976.2021.1894070

Rackaityte, E. and Lynch, S. (2020) 'The human microbiome in the 21st century.' *Nature Communications 11*, 5256. Accessed on 18/08/2021 at www.nature.com/articles/s41467-020-18983-8

Rediger, J. (2020) *Cured*. London: Penguin.

Reese, A.T., Madden, A.A., Joossens, M., Lacaze, G. *et al.* (2020) 'Influences of ingredients and bakers on the bacteria and fungi in sourdough starters and bread.' *mSphere 5*, 1. Accessed on 21/08/2021 at https://journals.asm.org/doi/full/10.1128/mSphere.00950-19

Riccio, P. and Rossano, R. (2020) 'The human gut microbiota is neither an organ nor a commensal.' *FEBS Letters 594*, 20, 3262–3271. Accessed on 21/08/2021 at https://febs.onlinelibrary.wiley.com/doi/10.1002/1873-3468.13946

Rogers, G.B., Keating, D.J., Young, R.L., Wong, M.L. *et al.* (2016) 'From gut dysbiosis to altered brain function and mental illness: Mechanisms and pathways.' *Molecular Psychiatry 21*, 6, 738–748. Accessed on 19/08/2021 at www.nature.com/articles/mp201650.pdf

Sadanand, S. (2019) 'Beyond bacteria: Studies of other host-associated microorganisms.' *Nature Portfolio*. Accessed on 18/08/2021 at www.nature.com/articles/d42859-019-00013-3

Salim, S.Y., Kaplan, G.G. and Madsen, K.L. (2014) 'Air pollution effects on the gut microbiota: A link between exposure and inflammatory disease.' *Gut Microbes 5*,

2, 215–219. Accessed on 22/08/2021 at www.ncbi.nlm.nih.gov/pmc/articles/PMC4063847

Savage, N. (2019) 'The complex relationship between drugs and the microbiome.' *Nature 577*, S10–S11. Accessed 18/08/2021 at www.nature.com/articles/d41586-020-00196-0

Schroeder, B.O. and Bäckhed, F. (2016) 'Signals from the gut microbiota to distant organs in physiology and disease.' *Nature Medicine 22*, 1079–1089. Accessed on 21/08/2021 at www.nature.com/articles/nm.4185

Scientist, The (2019) 'Exercise changes our gut microbes, but how isn't clear yet.' 15 August. Accessed on 22/08/2021 at www.the-scientist.com/news-opinion/exercise-changes-our-gut-microbes--but-how-isnt-yet-clear-66281

Sencio, V., Machado, M.G. and Trottein, F. (2021) 'The lung–gut axis during viral respiratory infections: The impact of gut dysbiosis on secondary disease outcomes.' *Mucosal Immunology 14*, 296–304. Accessed on 22/08/2021 at www.nature.com/articles/s41385-020-00361-8

Seo, D.O. and Holtzman, D.M. (2020) 'Gut microbiota: From the forgotten organ to a potential key player in the pathology of Alzheimer's Disease.' *The Journals of Gerontology Series A 75*, 7, 1232–1241. Accessed on 18/08/2021 at https://academic.oup.com/biomedgerontology/article/75/7/1232/5628721

Simon, J.C., Marchesi, J.R., Mougel, C. and Selosse, M.-A. (2019) 'Host–microbiota interactions: From holobiont theory to analysis.' *Microbiome 7*, 5. Accessed on 18/08/2021 at https://microbiomejournal.biomedcentral.com/articles/10.1186/s40168-019-0619-4

Slattery, C., Cotter, P.D. and O'Toole, P.W. (2019) 'Analysis of health benefits conferred by *Lactobacillus* species from kefir.' *Nutrients 1*, 11(6). Accessed on 21/08/2021 at www.mdpi.com/2072-6643/11/6/1252/htm

Stephenson, C. (2011) *The Complementary Therapist's Guide to Conventional Medicine.* London: Churchill Livingstone.

Takeshita, K. (2020) 'Sharpening the focus: Acupuncture interrupts the brain–gut vicious cycle underlying functional dyspepsia.' *Digestive Diseases and Sciences 65*, 1578–1580. Accessed on 27/08/2021 at https://link.springer.com/article/10.1007%2Fs10620-020-06080-1

Taneyo Saa, D., Di Silvestro, R., Dinelli, G. and Gianotti, A. (2017) 'Effect of sourdough fermentation and baking process severity on dietary fiber and phenolic compounds of immature wheat flour bread.' *LWT Food Science and Technology 83*. Accessed on 22/08/2021 at www.researchgate.net/publication/316531312_Effect_of_sourdough_fermentation_and_baking_process_severity_on_dietary_fibre_and_phenolic_compounds_of_immature_wheat_flour_bread

Umeda, N. (2019) 'Gut flora "the second brain" connects Eastern and Western medicine: Intestinal hyper-permeability or Qi deficiency can affect brain, mind, and whole body.' *Longhua Chinese Medicine 2*. Accessed on 18/08/2021 at https://lcm.amegroups.com/article/view/5146

Underwood, M., Mukhopadhyay, S., Lakshminrusimha, S. and Bevins, C. (2020) 'Neonatal intestinal dysbiosis.' *Journal of Perinatology 40*, 11, 1597–1608. Accessed on 18/08/2021 at www.nature.com/articles/s41372-020-00829-2

Voigt, R.M., Forsyth, C.B., Green, S.J., Engen, P.A. *et al.* (2016) 'Circadian rhythm and the gut microbiome.' *International Journal of Neurobiology 131*, 193–205.

Wang, H., Wang, Q., Liang, C., Su, M. *et al.* (2019) 'Acupuncture regulating gut microbiota in abdominal obese rats induced by high-fat diet.' *Evidence-Based Complementary and Alternative Medicine.* Accessed on 22/08/2021 at www. hindawi.com/journals/ecam/2019/4958294

Wassermann, B., Müller, H. and Berg, G. (2019) 'An apple a day: Which bacteria do we eat with organic and conventional apples?' *Frontiers in Microbiology 10*. Accessed on 21/08/2021 at www.frontiersin.org/articles/10.3389/fmicb.2019.01629/full

Weersma, R., Zhernakova, A. and Fu, J. (2020) 'Interaction between drugs and the gut microbiome.' *Gut 69*, 8. Accessed on 18/08/2021 at https://pubmed.ncbi. nlm.nih.gov/32409589

White, E. (2019) 'Impact of diet–microbiota interactions on human metabolism.' *Nature Portfolio.* Accessed on 18/08/2021 at www.nature.com/articles/ d42859-019-00017-z

Williams, M. (2005) 'Dietary supplements and sports performance: Amino acids.' *Journal of the International Society of Sports Nutrition 2*, 2, 63–67. Accessed on 19/08/2021 at https://jissn.biomedcentral.com/track/pdf/10.1186/1550-2783-2-2-63.pdf

Willmont, D. (2007) *Energetic Physiology.* Marshfield, MA: Willmountain Press.

Willyard, C. (2021) 'How gut microbes could drive brain disorders.' *Nature 590*, 22–25. Accessed on 22/08/2021 at www.nature.com/articles/d41586-021-00260-3

World Health Organization (2018) *Climate Change and Health.* Geneva: WHO. Accessed on 18/08/2021 at www.who.int/news-room/fact-sheets/detail/ climate-change-and-health

Worsley, J.R. (1998) *The Five Elements and the Officials.* J.R. and J.B. Worsley.

Xu, X., Feng, X., He, M., Zhang, Z. *et al.* (2020) 'The effect of acupuncture on tumor growth and gut microbiota in mice inoculated with osteosarcoma cells.' *Chinese Medicine 15*, 33. Accessed on 22/08/2021 at https://cmjournal.biomedcentral. com/articles/10.1186/s13020-020-00315-z

Yong, E. (2017) *I Contain Multitudes.* London: Vintage.

York, A. (2019) 'Microbiota succession in early life.' *Nature Portfolio.* Accessed on 18/08/2021 at www.nature.com/articles/d42859-019-00010-6

Subject Index

Author Index